# Island Shores, Distant Pasts

BIOARCHAEOLOGICAL INTERPRETATIONS OF THE HUMAN PAST:
LOCAL, REGIONAL, AND GLOBAL PERSPECTIVES

UNIVERSITY PRESS OF FLORIDA

Florida A&M University, Tallahassee
Florida Atlantic University, Boca Raton
Florida Gulf Coast University, Ft. Myers
Florida International University, Miami
Florida State University, Tallahassee
New College of Florida, Sarasota
University of Central Florida, Orlando
University of Florida, Gainesville
University of North Florida, Jacksonville
University of South Florida, Tampa
University of West Florida, Pensacola

# Island Shores, Distant Pasts

## Archaeological and Biological Approaches to the Pre-Columbian Settlement of the Caribbean

EDITED BY

SCOTT M. FITZPATRICK AND ANN H. ROSS

*Foreword by Clark Spencer Larsen*

University Press of Florida
Gainesville/Tallahassee/Tampa/Boca Raton
Pensacola/Orlando/Miami/Jacksonville/Ft. Myers/Sarasota

22  21  20  19  18  17   6  5  4  3  2  1

First cloth printing, 2010
First paperback printing, 2017

Library of Congress Cataloging-in-Publication Data
Island shores, distant pasts : archaeological and biological approaches to the pre-
Columbian settlement of the Caribbean / edited by Scott M. Fitzpatrick and Ann
H. Ross ; foreword by Clark Spencer Larsen.
p. cm.—(Bioarchaeological interpretations of the human past : local, regional, and
global perspectives)
Includes bibliographical references and index.
ISBN 978-0-8130-3522-2 (cloth: alk. paper)
ISBN 978-0-8130-5468-1 (pbk.)
    1. Indians of the West Indies—Antiquities. 2. West Indies—Antiquities.
3. Caribbean Area—Antiquities. 4. Island archaeology—Caribbean Area.
5. Ethnoarchaeology—Caribbean Area. 6. Human remains (Archaeology)—
Caribbean Area. 7. Human biology—Caribbean Area—History. 8. Human
settlements—Caribbean Area—History. 9. Land settlement—Caribbean
Area—History. 10. Human beings—Caribbean Area—Migrations—History.
I. Fitzpatrick, Scott M. II. Ross, Ann H.
F1619.I85 2010
972.9'01—dc22          2010020784

The University Press of Florida is the scholarly publishing agency for the State
University System of Florida, comprising Florida A&M University, Florida At-
lantic University, Florida Gulf Coast University, Florida International University,
Florida State University, New College of Florida, University of Central Florida,
University of Florida, University of North Florida, University of South Florida,
and University of West Florida.

University Press of Florida
15 Northwest 15th Street
Gainesville, FL 32611-2079
http://upress.ufl.edu

To Jim Petersen, one of the original participants in the session from which this volume is derived and whose untimely death left a major void in Caribbean archaeology.

# Contents

# Figures

# Tables

# Foreword

The settlement and colonization of the Caribbean—often referred to as the West Indies—has been a focus of study by anthropologists for more than a century. Questions that have framed much of the research in the region pertain to population origins and movement: Where did human population derive from? Did the migration into the region involve one or several migrations? When did humans first occupy the region? Building on the pioneering work of Irving Rouse, the archaeological community has been especially active in documenting the societies that once occupied the region, their material culture, timing of key economic and behavioral adaptations, and the living descendants. In the recent archaeological history of the region, the ambitious research programs of Kathleen Deagan, William Keegan, Samuel Wilson, Elizabeth Righter, Antonio Curet, Scott Fitzpatrick, Lee Newsom, Elizabeth Wing, and others have addressed a wide range of important issues that has given new understanding of this dynamic setting, its prehistoric inhabitants, and cultural dynamics that inform our understanding of past populations. Until recently, an obvious area missing from the discussion has been physical anthropology, with respect to the biology of both ancient and living populations. In part, this gap in the record is due to earlier misconceptions that human populations that inhabited the region over the past 10,000 years were not especially variable and that well-documented human remains from archaeological contexts were nonexistent. This picture is changing.

In recent years, study of living and past people in the region has helped to fill a void. By drawing upon the expertise of various areas of investigation, including both archaeology and physical anthropology, this new volume seeks to provide a collaborative, interdisciplinary perspective on the region by bringing together different lines of inquiry for developing a more comprehensive understanding of colonization and settlement. These chapters synthesize an important new body of evidence of what is currently known (and not known) about the archaeological and human biological past and present. New chronological data based on a wealth of radiocarbon evidence, computer modeling of population movement through the region, morphometric analysis of skeletal remains, stable isotope data, and exciting new developments in DNA variation give new insights into population and population movement. The discussions presented in this volume do not claim to be conclusive. Rather, collectively, they provide us with a means for promoting new discussions, new collaborations, and new understandings of the Caribbean and its people. Many questions remain, but the contributions to this volume represent a new step forward and an important basis for future study.

*Clark Spencer Larsen*
*Series Editor*

# Introduction

## Crossing the Caribbean Divide

### Integrating Anthropological Analyses in the Study
### of Pre-Columbian Cultures

SCOTT M. FITZPATRICK AND ANN H. ROSS

The Caribbean is the world's second-largest sea and seventh-largest body of water, encompassing an area of 2,754,000 square kilometers (1,063,000 square miles) and stretching 1,700 kilometers north–south from Florida to Panama and 2,300 kilometers east–west from the Antilles to the Yucatán Peninsula (figure 0.1). Geographically, the Caribbean comprises several island chains that are typically separated into three major groups—the Greater Antilles, Lesser Antilles, and Bahamas (including the Turks and Caicos). This is a general distinction, for other groups such as those adjacent to the South American mainland (for example, the "ABC" islands of Aruba, Bonaire, and Curaçao), the Caymans, and Trinidad and Tobago do not readily fit into these categories yet are important nonetheless for examining pre-Columbian settlement patterns and adaptations.

The larger islands of the Caribbean are a mixture of volcanic and limestone continental rock, while the Antillean chain is primarily composed of younger volcanic and coral islands. The Caribbean, because of its tropical climate, oceanography, and proximity to various physiographically distinct landmasses, is extremely diverse ecologically. It is home to 2.3 percent of the world's endemic plant species and 2.9 percent of endemic vertebrate species. These percentages are significant, considering that the Caribbean contributes only 0.15 percent of the Earth's surface. In addition, over 1,500 species of fish, 25 coral genera, more than 600 mollusk species, and numerous echinoderms, crustaceans, sea mammals, sponges, birds, and reptiles have been recorded in marine, freshwater, brackish, and terrestrial environments. The diversity of plant and animal taxa found in the Caribbean prehistorically and historically is well illustrated by Newsom

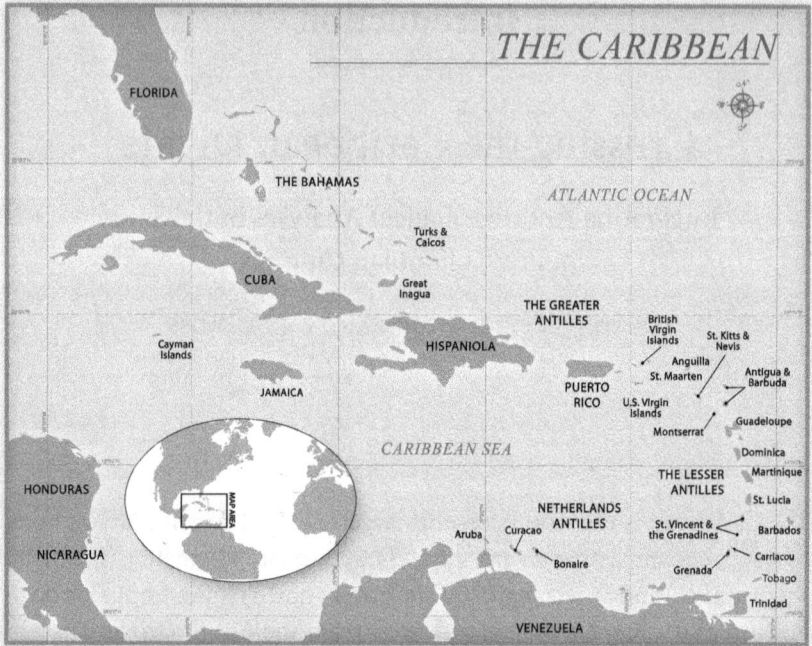

Figure 0.1. Map of the Caribbean (drafted by Michael Scisco, BioGeo Creations).

and Wing (2004) and deFrance and Newsom (2005). Despite the region's past and present ecological diversity, it is under tremendous threat from a variety of human-related events, including pollution, global warming, overfishing, and development, prompting Conservation International to designate the Caribbean as one of the world's 25 "Hotspots"—regions that are relatively small but contain high percentages of endemic species (see www.biodiversityhotspots.org).

For over a century, antiquarians, archaeologists, and anthropologists have searched for evidence of when and how peoples first settled the Caribbean islands (see Keegan 1994, 1996, 2000). Related questions addressing how Amerindians adapted to new insular landforms, where they began constructing permanent villages, and the mechanisms used for exploiting their surroundings have also been of great interest to scholars. Investigations of these lines of inquiry have helped researchers to better understand migratory patterns, group interactions on mainlands and islands, and the transformations of insular environments that were shaped to suit Caribbean peoples' particular needs, largely through agricultural activities and the associated rise in population.

The most enduring research questions related to the pre-Columbian Caribbean, however, have revolved around colonization events and settlement patterns. The main foci of these studies have often involved developing cultural typologies (*sensu* Rouse 1986, 1992a). More recent technological advances in the physical and natural sciences, such as radiocarbon dating and geochemical and mineralogical analyses, have greatly increased the ability of archaeologists to address many issues related to the peopling of these islands. Researchers in archaeology and other branches of anthropology, particularly skeletal biology and genetics, have also opened up new avenues of inquiry by using a host of sophisticated techniques, including stable isotope and DNA analyses. This has greatly diversified and expanded our efforts to track ancient population movements (for example, biodistance, demography, and affinal relationships) that complement existing archaeological data. Many of these advances have been successfully applied to the study of prehistoric population expansions in other geographic regions, and interest in conducting and integrating similar studies is growing in the Caribbean.

Caribbean archaeologists have been very proactive in developing cultural typologies and trying to decipher the various patterns of colonization and migration that occurred over the course of six thousand years or so. Several books published in the past twenty years, including Keegan's *People Who Discovered Columbus* (1992), Newsom and Wing's *On Land and Sea: Native American Uses of Biological Resources in the West Indies* (2004), Curet's *Caribbean Paleodemography: Population, Culture History, and Sociopolitical Processes in Ancient Puerto Rico* (2005), and Keegan's *Taíno Indian Myth and Practice: The Arrival of the Stranger King* (2007), have all shown the importance of integrating and interpreting decades of archaeological data. Edited volumes by Siegel (1989, 2005), Curet and colleagues (2005), and Delpuech and Hofman (2004), for example, have also improved our ability to bring together, and shown the necessity of bringing together, a wide range of scholars in archaeology to comment on and debate current issues in prehistory. There have been few attempts, however, to integrate the multiple perspectives within anthropology.

Wilson's edited volume, entitled *The Indigenous People of the Caribbean* (1997), was important because it is one of the few attempts to interweave perspectives from archaeology, cultural anthropology, linguistics, and ethnohistory to discuss early West Indian inhabitants and their ancestors who came into contact with Europeans. Wilson's *Archaeology of the Caribbean*

(2007), Newsom and Wing's (2004) book, and Hofman and colleagues' *Crossing the Borders: New Methods and Techniques in the Study of Archaeological Materials from the Caribbean* are more recent attempts to synthesize the growing corpus of archaeological data, following a more holistic approach to understanding pre-Columbian peoples in the region.

However, collaborative efforts in the Caribbean have really only now begun to reach the point where researchers can tackle issues of population expansion and mobility in a more systematic and comprehensive way. Archaeology has clearly provided the bulk of the evidence for explaining human settlement patterns during the pre-Columbian colonization of the Caribbean, but the data are severely limited in many respects, in part because of the tropical climate, which is generally not conducive to the long-term preservation of organic remains. Archaeological sites are also easily destroyed by development, erosion, and looting. As a result, archaeology provides us with only a portion of the larger puzzle of how pre-Columbian populations expanded into their island world. Geographical gaps in the research must also be considered. A heavy emphasis on archaeological surveys in the northern Antilles compared to the southern part (particularly those islands south of Martinique) leaves us with many unanswered questions of how Amerindian groups made their way through the Caribbean.

## Temporal Framework

The Caribbean, also commonly referred to as the West Indies, was initially colonized by two separate waves of hunter-gatherer-foraging groups. The first of these was probably to Trinidad and Tobago around 7000 BP; another migration, slightly later, may have taken place from Mesoamerica to Cuba and Hispaniola about 6000 BP (Keegan 1994, 2000). Both migrations occurred when sea levels were significantly lower than they are today, shortening the distance between landmasses and, in some places, exposing previously submerged islands. Two other major migrations occurred a few thousand years later and represent the initial colonizers of the Lesser Antilles and Puerto Rico. The first of these is usually termed the Archaic and happened around 4000–4500 BP, while a separate migration of ceramic-making horticulturalists known as Saladoid made their way into Puerto Rico, the Virgin Islands, and the Lesser Antilles around 2500 BP. These groups, after what appears to be a lengthy hiatus, eventually spread into the Greater Antilles and Bahamas between 1500 BP and 1300 BP, developing into a wide range of different cultural groups (Keegan 2000). Because of these geographically broad and temporally disparate conditions that led to several major (and probably multiple minor) migrations, the Caribbean presents some unique opportunities and challenges for researchers in attempting to explain how prehistoric peoples colonized these islands and how settlement patterns were structured after initial occupation.

Some of the earliest work dedicated to looking at the origins of Amerindian peoples in the Caribbean was conducted by Fewkes (1907, 1914), who ventured through Puerto Rico and the Lesser Antilles collecting artifacts for the Smithsonian Institution in Washington, D.C. Researchers in the early part of the twentieth century included such prominent figures as Jesse Fewkes (1907, 1914), Froleich Rainey (1940), Ricardo Alegría (1965), Irving Rouse (1986, 1992a), and Ripley Bullen (Bullen and Bullen 1972), as well as Estrella Rey Betancourt, who all began to offer more detailed glimpses of the ancient Caribbean by examining and in many cases excavating sites in the Caribbean and on the South American mainland. Through these efforts, we have slowly begun to piece together when and how peoples colonized these islands, how cultural differences began manifesting themselves, and how these and other events structured settlement patterns, ultimately influencing resource use and social behavior through time (see Fitzpatrick and Keegan 2007).

The results of these investigations demonstrate that pre-Columbian peoples were interacting frequently (Hofman et al. 2008), adapting to variations in environment and resource availability, overexploiting many terrestrial and marine foods, and developing unique cultural traits. For a more in-depth treatment, readers can consult several key sources that summarize what we presently know about cultural periods in the Caribbean during pre-Columbian times: Rouse 1986, 1992a; Siegel 1989, 2005; Keegan 1994, 1996, 2000, 2007; Wilson 1997, 2007; Petersen et al. 2004; Newsom and Wing 2004; Fitzpatrick and Keegan 2007 (see figure 0.2 for a synopsis of cultural periods identified in the Caribbean).

## Biological Perspectives

The origins of Native Americans, their dispersal patterns throughout the New World, and the unique cultures that developed have also been a topic of fundamental interest to anthropologists and other scholars for centuries. However, discussions and studies addressing Native American origins

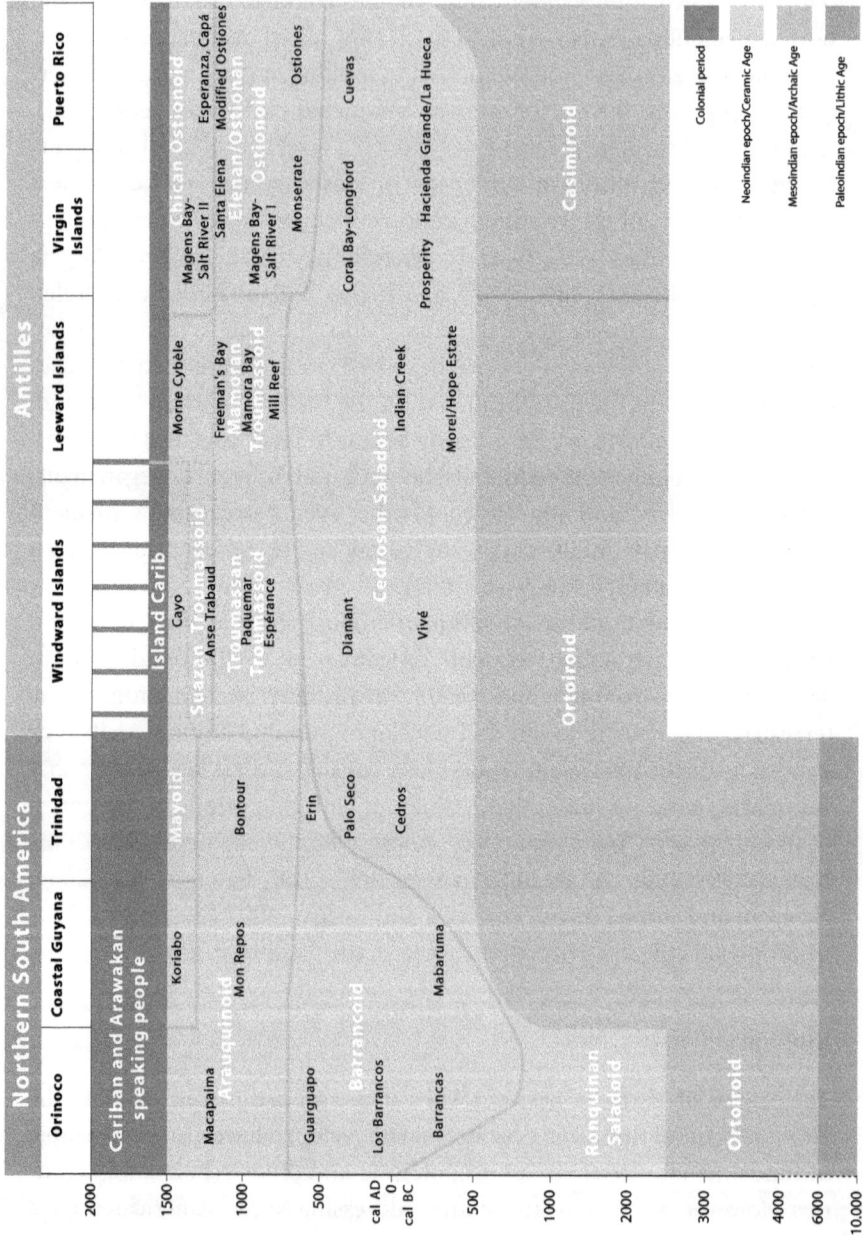

Figure 0.2. Cultural-historical model of pre-Columbian cultures in the Caribbean (reprinted with permission from C. L. Hofman).

**Northern South America**

| Orinoco | Coastal Guyana | Trinidad |
|---|---|---|

**Antilles**

| Windward Islands | Leeward Islands | Virgin Islands | Puerto Rico |
|---|---|---|---|

Cariban and Arawakan speaking people

Island Carib

Mayoid

Arauquinoid — Koriabo, Macapaima, Mon Repos

Barrancoid — Guarguapo, Los Barrancos, Barrancas, Mabaruma

Bontour, Erin, Palo Seco, Cedros

Suazan Troumassoid — Cayo, Anse Trabaud

Troumassan Troumassoid — Paquemar, Esperance

Morne Cybèle

Freeman's Bay, Mamora Bay

Mamoran Troumassoid — Mill Reef

Cedrosan Saladoid — Diamant, Vivé, Indian Creek, Morel/Hope Estate

Chican Ostionoid — Magens Bay-Salt River II, Santa Elena

Esperanza, Capá, Modified Ostiones

Elenan/Ostionan Ostionoid — Magens Bay-Salt River I, Monserrate

Ostiones

Coral Bay-Longford, Cuevas, Prosperity, Hacienda Grande/La Hueca

Ronquinan Saladoid

Ortoiroid

Casimiroid

Ortoiroid

Legend:
- Colonial period
- Neoindian epoch/Ceramic Age
- Mesoindian epoch/Archaic Age
- Paleoindian epoch/Lithic Age

Timescale: 2000, 1500, 1000, 500, cal AD, cal BC, 500, 1000, 2000, 3000, 4000, 6000, 10.000

have not been without disagreement or even controversy. In the United States, for example, the study of Native American populations has recently received a great deal of attention with the very public court proceedings surrounding the "Kennewick Man" *(Bonnichsen et al. v. United States)* case, whose outcome has had a profound academic impact as well as political ramifications for the study of Native Americans.

Two key objectives in physical anthropology are to document the vast range of human variation of past and present populations and to investigate the evolutionary and environmental forces responsible for phenotypic variation, which then allows one to address large-scale issues such as migration and expansion. The patterns of human variation among geographic populations have been examined using genetic markers, linguistics, and anthropometrics. In particular, much attention has been directed at investigating genotypic and phenotypic variation among past and present New World populations.

Historically, the Americas have been viewed as relatively homogenous without significant biological variability until initial European contact and the subsequent influx of European and African populations, a view profoundly influenced by Morton and Hrdlička (Powell and Neves 1999). This notion of negligible population variability in the Americas, and thus lack of antiquity, can be found at the core of Americanist studies until recently (Ross et al. 2002a). Recent investigators have focused on revisiting many of the old "typological" paradigms presented by previous researchers and have advanced the field by utilizing new models of human migratory patterns, although there is still no consensus with respect to past cranial diversity (Jantz and Owsley 2001; Lahr 1995; Powell and Neves 1999).

Researchers have generally concentrated on localized and regionalized studies, in particular, South American and South Andean variation (Bernal et al. 2007; Cocilovo and Rothhammer 1990, 1996, 1999; Fabra et al. 2007; Perez et al. 2007; Ross et al. 2008; Rothhammer et al. 1982, 1984, 1985; Varela and Cocilovo 1999, 2000, 2002). Population origins traced to early demic expansion represent one favored explanation for Native American homogeneity (Hrdlička 1920; Turner 1985, 1994). Linguistic evidence is another area of exploration used in the study of early American migration models. Greenberg (1987) and colleagues (Greenberg and Ruhlen 1992; Greenberg et al. 1986) suggest that all living Amerindians are direct descendants of a single founding population. The assumption is that a common language indicates a common origin that can act as a

barrier to gene flow, which influences the formation of population structures approximating geographic isolation (Barbujani and Sokal 1990; Sokal
et al. 1993). Alternatively, multiple migration models have been presented
by linguistic, genetic, dental, and craniometric studies of late Pleistocene
and early Holocene Paleoindians (Haydenblit 1996; Lahr 1995; Powell and
Neves 1999; Neves and Pucciarelli 1991; Steele and Powell 1993, 1994; Jantz
and Owsley 2001; González-José et al. 2005). However, each of the above
models requires that a priori assumptions be made about the amount of
phenotypic variation present in the Americas before European contact. In
a recent study, Ross and colleagues (2002a) found morphological similarities between a precontact Mexican and a coastal Ecuadorian sample and
dissimilarity to the Howells Peruvian and Cuban samples, which contra-
dicts conclusions by Ruhlen (1994) and others that South America was
populated by a single migration (see also Callaghan 2003a). Since craniofacial morphological similarities reflect genetic relationships to some
degree, one can further extrapolate that the morphological similarity between Mexican and Ecuadorian crania may be the result of early demic expansion concurring with the archaeological evidence of contact between
Mexico and Ecuador (Ross et al. 2002a; Ubelaker 1987). In a recent study,
Ross and colleagues (2008) found a significant difference between coastal
and highland precontact Peruvian populations, which they attributed to
both local and altitudinal adaptations following initial settlement into
South America, followed by subsequent genetic isolation, with the Andes
Cordillera acting as a genetic barrier. These results provide further support
for the argument that different populations colonized the New World and
emphasize the need for continued investigations (Neves et al. 2007; Schurr
et al. 1999; Schurr and Wallace 1999; Schurr and Sherry 2004).

Until recently, the study of population diffusion into the Caribbean
was virtually ignored by physical anthropologists and geneticists alike
(Coppa et al. 2002; Lalueza-Fox et al. 2001, 2003; Martínez-Cruzado 2002;
Martínez-Cruzado et al. 2001, 2005; Toro-Labrador et al. 2003; Ross et al.
2002a; Ross 2004). In part, this is probably due to the scarcity of available well-preserved skeletal material and perhaps a bias toward investigating the settlement of Paleoamericans with little interest given to later
population migrations. Before conclusions or interpretations can be made
with respect to early Paleoindian and subsequent population variation
and/or migration models (or the effects of European and African populations on Latin American populations), we must first investigate the spatial

variation present before European contact and after the period of major Native American population migrations into the New World, including the Caribbean.

## The Rationale for This Volume

Although a century of work by antiquarians and both professional and avocational archaeologists has greatly improved our understanding of how and when humans migrated to islands in the Caribbean, it is clear that we are often hampered by the poor preservation of organic remains, geographical gaps in research, and many other issues. If we are to truly decipher how pre-Columbian peoples moved between and interacted with other islands, then we must collate our findings with the increasing amount of biological data from genetics and skeletal biology. Only then can we hope to begin to appreciate the magnitude of these island shores over such distant pasts.

An increasing number of archaeological projects, both research focused and contract driven, have highlighted areas that deserve greater attention (for example, see Curet et al. 2005). Fortunately, a recent surge of interest by biological anthropologists has paved a new road for tracking ancient and modern population movements in the circum-Caribbean (Coppa et al. 2002; Lalueza-Fox et al. 2001, 2003; Martínez-Cruzado 2002; Martínez-Cruzado et al. 2001, 2005; Ross 2004; Toro-Labrador et al. 2003). We have now reached a stage where contributions from numerous scholars within anthropology, not just archaeologists, can incrementally expand our ability to trace the movement of human populations into and within the Caribbean islands.

As such, we felt it was an opportune time to bring together a group of anthropologists and archaeologists who were interested in examining issues related to the colonization and settlement of the pre-Columbian Caribbean. To further this effort, we organized a special session at the 71st Society for American Archaeology conference in San Juan, Puerto Rico, entitled "New Perspectives on the Prehistoric Settlement of the Caribbean." We were extremely fortunate in receiving commitments from a number of established and rising scholars who resided at institutions in Antigua, Puerto Rico, Canada, the Netherlands, the United Kingdom, and the United States. This was an especially appropriate venue to present and discuss some of the more innovative and cutting-edge research projects currently under way: not only was the location Caribbean-based but also the conference was one of the most well-attended meetings in SAA history.

In developing the session, there were, of course, many trajectories we could have taken. After much discussion, we chose to focus on issues surrounding how and when islands in the Caribbean were settled in the past, the movement of these populations over time, and how we could identify these phenomena using archaeological and biological data. Even by narrowing the session and the volume to these particular realms of inquiry, we began to unfold a number of different subjects that have been, or are currently, at the forefront of archaeological and anthropological discourse, which would benefit from inter- and intra-subdisciplinary contact.

The chapters in this volume are an attempt to synthesize many facets of what we know archaeologically and biologically about the pre-Columbian settlement of the Caribbean and to highlight the various data sources at our disposal. These include a heavy emphasis on radiocarbon dating (chapters 2, 4, and 8), which has largely been under- or misused in the Caribbean to establish cultural chronologies (see Fitzpatrick 2006), computer simulations of seafaring (chapter 6), three-dimensional and traditional craniometrics (chapter 5), stable isotopes (chapter 7), and mitochondrial and nuclear DNA analysis (chapters 3 and 9). As is demonstrated in the following chapters, these techniques show great promise for helping us to address issues of pre-Columbian Caribbean population expansions and demonstrate the utility of integrating and comparing biological markers with the archaeological record. We optimistically anticipate that the end result will promote further discussion and collaborative investigations on the origins and movement of ancient Amerindians, helping us to resolve longstanding questions that we could not hope to answer alone in a region that has, to date, received comparatively little attention compared to other parts of the New World.

We would like to thank all of those who participated in the session and those who included their work here. Special thanks also go to John Byram, who helped us through each stage of the production process, and to the reviewers and editorial staff at the University Press of Florida who gave critical comments on the chapters and organization of the book.

# 1

# Island Shores and "Long Pauses"

WILLIAM F. KEEGAN

## Island Shores

Antonio Curet (2005) recently noted that surprisingly little specific and systematic attention has been paid to migrations, population movements, and island colonization in the Caribbean islands. To a large degree, Caribbean archaeologists have simply followed Irving Rouse's (1986) conclusion that there were four separate migrations that announced the beginnings of the Lithic, Archaic, Ceramic, and Historic ages. The one notable exception is the general framework for island colonization proposed by Keegan (1985, 1995, 2004; Keegan and Diamond 1987; Siegel 1991). Yet very few specific studies of the movement of peoples through the islands have been initiated.

The chapters in this volume reflect significant new efforts to correct that deficiency. A major issue concerns the timing of migrations. According to the time-space, culture-historical framework proposed by Rouse (1992a), the beginning of each age was simultaneous throughout the islands. Thus, a date of 400 BC for the Ceramic Age settlement of Puerto Rico is taken as evidence for the equally early settlement of the Lesser Antilles. Yet we now know that most of the Lesser Antilles were not settled until about AD 200 (see chapters 2 and 8, this volume) and that most, if not all, of the Lesser Antilles were bypassed during the early Ceramic Age (Keegan 2004). New radiocarbon dates are beginning to provide clarity to this issue, but we must remain mindful of the corrections, calibration, and standard deviations associated with these dates (Davis 1988; Fitzpatrick 2006).

Moreover, basing conclusions about population movements on the distribution of material remains can be a risky endeavor. Rouse (1992a) concluded that similarities between ground-stone tools (the hallmark of the Archaic Age) in the Greater Antilles and those discovered at the Banwari Trace site in Trinidad provided evidence for the movement of peoples from the latter to the former through the intervening Lesser Antilles. Richard Callaghan (chapter 6) develops an alternative perspective on the origins of so-called Archaic peoples: he proposes that there was no Archaic migration through the Lesser Antilles. In fact, the adoption of ground-stone tools may have derived from independent invention or the diffusion of this technology from the mainland, and not a separate migration of people.

In a similar vein, more detailed investigations of the entire assemblage of archaeological materials are needed to clarify both the large-scale movements of peoples and the smaller-scale interactions among different groups (see chapters 2 and 7, this volume). Of special note is Menno Hoogland and Corinne Hofman's use of strontium isotopes to investigate post-mortem mobility in burial populations (chapter 7; also see Keegan 2009).

Skeletal remains provide additional sources of valuable information. To his credit, Rouse (1992a) recognized that physical (biological) characteristics are important for defining "peoples and cultures." With the exception of the chapters in this volume, few archaeologists have addressed this issue. The use of craniometrics to assess variability in ancient Caribbean populations promises important new insights (see chapter 5, this volume). Beginning from the recognition that the earliest Lithic Age colonists preceded Ceramic Age colonists by thousands of years and that they came from completely different populations, we would expect to observe biometric differences (Coppa et al. 2008). Add to this that some groups practiced cranial deformation, and the potential for craniometric differences increases dramatically. The main constraint at this time is the relatively small sample size available for analysis.

The investigation of mitochondrial DNA (mtDNA) offers an additional line of evidence for identifying differences in human populations. At present, interpretations based on these data also are limited by small sample sizes, although the results do suggest that the islands were colonized by at least two distinct groups (see chapters 3 and 9, this volume). However, mtDNA provides only half of the answer. Recent studies of the Y chromosome have demonstrated that in at least some episodes of colonization, the males from one population interbred with females from a different population (for example, Chikhi et al. 2002; Hage and Marck 2003). Genetic evidence for both males and females will be necessary to accurately reconstruct the dispersal of different populations.

The chapters in this volume report detailed and systematic efforts to use time, material culture, and human biology to elucidate the details of island colonization. Yet the colonization of islands by humans involves the movement of social groups. Social relations are the focus of the remainder of this chapter, which originally was prepared for an SAA symposium that focused on the contributions of Jim Hill to American archaeology. Jim is best known as one of the pioneers of "ceramic sociology," in which ceramic decorations were used to identify social groups (Deetz 1968; Hill 1970; Longacre 1970; compare Roe 1981). He went on to investigate equilibrium models of prehistoric change (Hill 1977) and the role of the individual in material culture patterning (Hill and Gunn 1977). Although I had only one graduate seminar with Jim at UCLA, I was surrounded by his students for four years and thus received elements of Jim's wisdom on an almost daily basis. Jim Hill left a lasting impression on me. His legacy is reflected in my efforts to use settlement patterns to investigate social organization in the Caribbean islands, and I later extended these investigations to Polynesia in collaboration with Per Hage (University of Utah). Per pioneered the use of mathematically derived structural models in anthropology (Hage and Harary 1983, 1996) and investigated Oceanic social organization from linguistic (Hage 1998, 1999) and Y-chromosome DNA (Hage and Marck 2003) perspectives. A former student of his put us in contact, but unfortunately we never met and were unable to complete our collaborative effort before his untimely death. This chapter is written in memory of both Jim and Per.

From Ceramic Sociology to Settlement Pattern Sociology

Up until the late 1960s, the prevailing view was that prehistoric social organization could not be detected through archaeological analysis. However, Hill (1970), Deetz (1968), and Longacre (1970) showed in their studies of ceramic decoration that material culture could reflect elements of social organization. In doing so, they opened a new vista for archaeologists. From their work I came to understand that social organization could be studied but that we needed to do so in a rigorous and carefully constructed way.

My dissertation research examined settlement patterns in the prehistoric Bahamas. I found that 90 percent of the archaeological sites occur in pairs, with pairs defined as sites less than five kilometers apart. Every pair typically was separated from the next pair by at least ten kilometers. Morgan Maclachlan and I sought an answer to this patterning, and we concluded that these pairs were the product of social, rather than economic, factors. In a paper published in *American Anthropologist* (Keegan and Maclachlan 1989), we developed a model for the emergence of avunculocal chiefdoms among the Taínos. Our model is not without its detractors. One issue is whether the sites in these pairs were contemporaneous or sequential. Although very little research has been conducted in the Bahama archipelago, two pairs have been shown to be contemporaneous and none have been shown to be sequential.

The distances (five versus ten kilometers) may not seem particularly significant. However, the reality of community identity on a small island was brought home to me during fieldwork on Middle Caicos in the Turks & Caicos Islands. There are three settlements on this eighty-square-kilometer island, and they are located about ten kilometers apart. The settlements and people are not very different in their appearance and show few signs of difference in occupation, status, or wealth. Conch Bar is located near the western end and because the airport is located there, residents express a sense of superiority with regard to the other two settlements. Bambarra is located near the middle of the island, and the people there are viewed as country folk; Lorimers is at the far eastern end of the island, and the people there are viewed as "hillbillies." People from these communities intermarry and interact on an almost daily basis. Still, they associate themselves with their natal community. Every time I talk with Gertie Forbes, she reminds me that she is a Lorimers woman, and she complains that her husband, Simon, moved her to Bambarra. They have been married for at least fifty years, and she has lived in Bambarra for most of that time. Nevertheless, she always reminds me that she is from Lorimers, which is only ten kilometers from Bambarra.

Island Archaeology

It recently has been suggested that there is no need for a subdiscipline called "island archaeology" (Rainbird 1999; Boomert and Bright 2007). I disagree. Archaeologists who work on islands have been comparing notes for years. The cross-fertilization has provided important insights and led to the collaborative effort between Per Hage and myself with regard to the "long pauses" in the Caribbean and Polynesia. Tragically, Per died before we were able to complete the project.

There have been shortcomings in the broader study of islands. For example, island archaeologists have tended to emphasize geographic distance—the linear distance of water passages separating islands. This emphasis is a legacy of MacArthur and Wilson's "Theory of Island Biogeography" (1967), which focused on distance and area effects and other biological processes. Yet there are other ways to measure distance.

*Economic distance* reflects the boundaries of the area required to extract needed resources, which has often been expressed in terms of catchment analysis.

*Demographic distance* is the minimal distance required to maintain a reproductively viable population. To some degree, this is reflected in Pat Kirch's (2000) "lifeline" model, in which an offspring community maintains ties to the parent community. John Moore (2001) recently demonstrated by computer simulation that the viability of a colony was greatly enhanced by maintaining ties to one other settlement from which spouses could be recruited. In other words, relations between parent and offspring communities made reproductive survival much more likely.

*Ceremonial distance* provides yet another dimension. Ceremonial activities serve to reinforce the sense of group identity and to legitimize positions of leadership. What is the maximum geographical distance over which members of a "ceremonial group" can successfully be maintained?

*Political distance* raises the question of how far leaders can extend their influence beyond the village in which they live. In Hawai'i the "king" was peripatetic and moved between villages during the year.

There also is *social distance:* the maximum distance between settlements that still allows the members of a social group to participate in their social obligations.

All of those dimensions of distance are interrelated and codetermined (Keegan 2004). The point is that simply measuring a water passage, considering the water transport available, and then deciding on the probabilities for island colonization and long-term survival is not enough. A variety of factors influenced the ability of groups to assemble colonists who could achieve economic, demographic, ceremonial, political, and social success—what some might lump under the heading "social reproduction."

## Social Processes, the Forgotten Factor

A second tendency has been to treat colonizing groups as objects. This approach is clear in the present volume in studies of genetic and craniometric markers that are used to "source" human groups in much the same way as we source lithics. Moreover, demographic models have been used to create the impression that when population density reaches some critical level the group will simply fission to create one or more new colonies (Curet 2005; Keegan 1995; cf. Keegan 2004, 2010). We track the distribution of sites in time and space but pay little attention to the people who actually lived at these settlements.

Napoleon Chagnon's (1984) study of fissioning in a Yanomamo community showed that this was not a simple social process and that the splintering of a group might be described as gut-wrenching. Villages are composed of complex and chaotic webs of social relations. Before a new colony can be established, a leader must assemble not only the materials needed to establish a successful settlement but also a retinue of followers who fulfill the needs of social reproduction. Every individual in the colonizing group faces different, competing, and often conflicting goals. The decision to strike out is rarely an easy one.

One way to investigate social processes is to look at elements of social organization (for example, Murdock 1949). Although there is a danger in overformalizing social practices (Curet 2002; Goodenough 1955; Maclachlan and Keegan 1990), organizational differences provide important clues as to how colonizing groups were assembled, their composition, and the distances over which different groups might be successful. The role of social relations cannot be denied. We will never understand the colonial enterprise until we recognize colonists as socially organized groups of people.

## "Long Pauses" in the Caribbean and Polynesia

### The Caribbean

The colonization of the Caribbean has proved far more fascinating than originally described. The prevailing scheme recognizes a Lithic Age colonization of Cuba, Hispaniola, and Puerto Rico around 4000 BC, with colonists arriving from the Yucatán Peninsula and/or Belize; an Archaic Age that begins with the arrival of colonists from South America around 2500

BC; and a Ceramic Age (Saladoid) that commences with colonists from South America around 500 BC (Rouse 1992a). The last group is my main concern, but it must be remembered that there were other people living in the islands when they arrived.

The Caribbean is composed of classic, stepping-stone archipelagoes. Although Grenada is not visible from the South American mainland, after Grenada is reached there are islands visible to the north all the way to the Anegada Passage between the British Virgin Islands and Puerto Rico (but this too is a relatively short water gap). Conventional wisdom would suggest that every island in the Lesser Antilles was settled during a northward migration to the Greater Antilles. However, present evidence suggests that most of the islands in the Lesser Antilles were bypassed at this time (see chapters 2 and 8), although some researchers have argued that we simply have not found the early sites on these islands. A similar argument has been made for Polynesia.

The Saladoid has been characterized by a small percentage of artifacts that share identical iconography all the way from South America to Puerto Rico. I have described this situation as a "veneer" that served to create a sense of unity and group identity among peoples who were widely scattered through the islands at this time (Keegan 2004). A number of commentators have expressed difficulty in understanding the term *veneer*. On reflection, I now think it may be more appropriate to describe the elaborately decorated pottery with its symbolic meanings combined with the widely traded stone and shell beads and pendants as the material manifestations of a "lifeline." These objects reflect a shared identity and promised cooperation and support in response to the vagaries of island colonization faced by small and relatively isolated groups. After a thousand years, during which most of the Lesser Antilles were colonized and the population of the islands increased significantly, the Saladoid mode of expression disappeared. It was replaced by more local modes of expression, albeit still regional interaction spheres. As these developed, the Saladoid lifeline became superfluous.

Perhaps the strangest aspect of the Saladoid is that it stopped in Puerto Rico. The eastern Dominican Republic is literally a stone's throw away. Moreover, the prevailing framework holds that pottery and agriculture were introduced to the islands by Saladoid peoples and were not brought to Hispaniola, Cuba, Jamaica, and the Bahamas until the subsequent Ostionoid expansion began after AD 600. The Ostionoid has been portrayed as a degeneration of Saladoid pottery that occurred after the latter had been living on Puerto Rico for nearly a thousand years. But why was there this "long pause" in Puerto Rico, especially when the Dominican Republic was so close? The current answer is that Archaic peoples who already were living on Hispaniola prevented the Saladoids from settling on their island, although this answer is not entirely satisfying.

When Per Hage and I began our collaboration over five years ago, we focused on the prevailing model that there was an Ostionoid expansion from Puerto Rico that introduced pottery to the rest of the Greater Antilles and Bahamas. This scenario suggested that there might be parallels between the Caribbean and Polynesia, even if the distances between islands were substantially different. However, the situation has changed in the past five years. We now know that so-called Archaic peoples in Cuba, Hispaniola, and Puerto Rico were making pottery vessels up to two thousand years before the Saladoids arrived in Puerto Rico (and probably were practicing plant management if not true cultivation). Moreover, we suspect that they had developed complex social and political organizations, as did hunter-gatherer societies who rely on marine resources in California and Florida, and that the introduction of new cultigens from South America and the need to repel the advances of the Saladoid intruders led to their further consolidation (Keegan 2006a; Rodríguez Ramos et al. 2008). The new perspective is that Ostionoid reflects not a migration out of Puerto Rico but rather an assimilation of Puerto Rico in an interaction sphere based in Hispaniola.

The social implications are not immediately apparent, but as Antonio Curet (2003: 19) notes with regard to the ethnohistoric Taíno peoples, "Hispaniolan and Puerto Rican polities used significantly different ideological foundations, a reflection of differences in the nature of the political structure and organizations." He continues, "Judging from the striking differences mentioned, they likely developed from distinct types of ancestral societies, and/or through different and divergent historical processes." In this case, the long pause seems to reflect a standoff between well-established indigenous (Archaic) peoples and the new arrivals. It appears that the Archaic peoples won, although their culture was transformed through their interactions with the Saladoids.

## Polynesia

It has long been recognized that there is a period of about 1,500 years between the settlement of western Polynesia and the colonization of eastern

Polynesia. This time gap has been the subject of numerous publications and heated debate. Some archaeologists feel that the long pause is real, while others think it is an artifact of the available evidence. The latter suggest that early sites eventually will be found in eastern Polynesia. They may be right, and I do not want to become embroiled in this debate. Instead, Per Hage and I assumed that the long pause was real, and we then looked for reasons why it may have occurred.

Oceania today has a bewildering variety of social institutions. Per Hage has tried to make sense of these and to retrodict ancestral patterns (Hage 1998, 1999). One of his conclusions is that the Lapita social organization was based on matrilineal institutions, not the patrilineal conical clan (Sahlins's *ramage*) by which it is more popularly known. This conclusion is significant with regard to social distance in the colonization of Oceania. Matrilineal groups tend to have short marriage distances because it is necessary for men who move to live in the village of their wife's family to maintain close ties to the village of their clan. Although related women form the domestic core of these villages, men typically occupy positions of leadership and must assemble for trading and even raiding expeditions. The question is, how far can men be isolated from their clan's village before they are unable to participate fully in their clan obligations?

Our conclusion is that the geographical distances between islands as far east as western Polynesia did not constrain male participation in clan activities. Eastern Polynesia is different. The distances between islands or archipelagoes are far greater. And while we agree with Geoff Irwin (1992) that return voyages were made, the distances that had to be crossed limited effective communication and participation with the homeland. The "long pause" (if it actually occurred) likely reflects numerous factors. These may include the improvement of vessels, navigation, storage of comestibles, and various other material constraints to long voyages.

Nevertheless, we suggest that the main factor was a change in social organization that facilitated the assembly of a group of colonists who could survive, for the most part, independent of the parent community. The transition from a matrilineal organization (even if it too was a conical clan) to a patrilineal conical clan made this possible. In a patrilineal conical clan, men were no longer widely dispersed in the villages of their wives; they now formed the core of the community and needed only to attract wives from other communities who were willing (or forced) to participate in these long-distance migrations. Others have addressed attributes of the patrilineal conical clan that make it an effective organization for the colonization of eastern Polynesia (for example, Kirch 2000).

In sum, we propose that the matrilineal foundations of Lapita cultures lacked the structural components required to assemble the facilities and retinue of colonists needed to settle the distant islands of eastern Polynesia. It was only after about 1,500 years of social and cultural development in western Polynesia that this situation changed, and the shift to a patrilineal conical clan made such long-distance voyages of colonization possible. I should also note that based on the distribution of different social systems in western Polynesia, Hage concluded that Samoa was the most likely source of these colonists. Because Per was the Pacific specialist, I leave it to my colleagues in the Pacific to judge the value of our suggestions.

## Conclusions

Per Hage showed me that questioning the conventional wisdom could lead to new and exciting interpretations. He was the consummate scientist who tested his ideas with cross-cultural evidence (Hage and Harary 1983, 1996; Hage and Marck 2003). Jim Hill taught me three things: first, that it was all right to make assumptions but that these needed to be stated as hypotheses with test implications that had to be rigorously tested; second, that archaeologists could indeed retrodict social organization based on material remains; and, perhaps most important, that sometimes you learn more from being wrong than from being right. Hopefully, I have been mostly right in this paper, but as I tell my students, if I had all the answers, I would write the book and we could all become lawyers.

# 2

# Rethinking Time
# in Caribbean Archaeology

## The Puerto Rico Case Study

RENIEL RODRÍGUEZ RAMOS, JOSHUA M. TORRES,
AND JOSÉ R. OLIVER

## Abstract

A radiocarbon database from Puerto Rico that includes more than five hundred dates is used to evaluate Rouse's chrono-cultural model for the island and the use of the "typological" form of time embedded in such a framework. To illustrate the inherent problems with the use of "typological" time, we present three cases: the Pre-Arawak/Saladoid-Huecoid interface, the "La Hueca problem," and the shifts in settlement patterns and the organization of communities registered during the Ceramic Age in the south-central portion of the island. These demonstrate that the application of "typological" time is inappropriate for addressing issues of varying scales of temporality and that it blurs the horizontal and vertical variability that existed in the island through time. The review of the available radiocarbon assays also indicates that none of the styles—and, for that matter, subseries, series, and periods—defined by Rouse for the island adheres to the temporality assumed in his most recent model. Therefore, these recently generated dates for Puerto Rico illustrate the need of developing similar databases in other islands, as it is quite possible that these will show a markedly different culture-historical panorama from that originally portrayed in Rouse's model.

## Resumen

Un inventario de fechamientos radiocarbónicos de Puerto Rico que incluye más de quinientas entradas es empleado para evaluar el modelo crono-cultural de Rouse para la isla y el uso de tiempo "tipológico" asumido en dicho esquema. Con el propósito de ilustrar los problemas inherentes con el uso del tiempo tipológico, examinamos tres casos: la interfase entre los grupos Pre-Arahuacos y las sociedades Saladoide y Huecoide, el "problema La Hueca," y los cambios en patrones de asentamiento y organización de comunidades de la Edad Cerámica en la porción sur-central de Puerto Rico. Estos casos demuestran que el empleo del tiempo "tipológico" es inapropiado para abordar asuntos de diversas escalas temporales y que también ofusca la variabilidad horizontal y vertical que existió en la isla a través del tiempo. La revisión de los fechamientos disponibles también indica que ninguno de los estilos y, por tanto, subseries, series, o periodos definidos por Rouse para la isla reflejan la temporalidad asumida en su modelo más reciente. De tal forma, estas fechas generadas recientemente en Puerto Rico ilustran la necesidad de desarrollar inventarios similares en otras islas ya que es muy probable que éstos presenten un panorama histórico-cultural marcadamente diferente al publicado originalmente en el esquema de Rouse.

*     *     *

Time has been the most underused and misused dimension in Caribbean archaeology. Since the establishment of the pottery chronology devised by the late Irving Rouse (1952, 1992a, 1992b; Rouse and Allaire 1978), most of us have assumed its premises in a quasi-religious fashion, merging culture and society into a single domain and considering that these changed con-comitantly along a unilinear temporal vector. Even though Rouse (1992b: 2) himself cautioned that "chronological charts are working hypotheses, intended to be revised as new data become available," since the initial con-struction of his cultural sequence for the Antilles nearly a half century ago, archaeologists have not made concerted efforts to either revise or refine it. Furthermore, we have not reconsidered the original temporal estimations and phylogenetic relationships for each of the cultural manifestations or paid adequate attention to the "typological" form of time (Fabian 1983: 23–24) that is embedded in such a model. The encapsulation of time in the squares of Rouse's chrono-cultural scheme becomes more problematic

| DATE | PERIOD | SERIES | SUBSERIES | | COMPLEX/STYLE | |
|---|---|---|---|---|---|---|
| | | | | | West | East |
| .D. 1200 - A.D.1500 | IVa | Ostionoid | Chican | | Boca Chica/Cap | Esperanza |
| ..D. 900 - A.D.1200 | IIIb | | Ostionan WEST | Elenan EAST | Late (Modified) | Santa Elena |
| ..D. 600 - A.D. 900 | IIIa | | | | Early (Pure) | Monserrate |
| ..D. 400 - A.D.600 | IIb | Saladoid | Cedrosan | | Cuevas | |
| 400 B.C. - A.D. 200 | | | | | | La Hueca* |
| 400 B.C. - A.D.400 | IIa | | | | Hacienda Grande | Hacienda Grande |
| 000 B.C. - 400 B.C. | I | Ortoiroid** | Corosan | | Coroso | |

Figure 2.1. Rouse's chrono-cultural sequence for Puerto Rico (adapted from Rouse 1992a).

* Huecan Saladoid subentries.

** Rouse (1992a: 67–68) argued that the Casimiroids of Hispaniola might have spread to western Puerto Rico where they established a frontier with the Corosans, as noted for instance in the Cerrillos site. However, in his chart he does not include Puerto Rico within the spatial extent of the Casimiroid series.

when those squares are used for addressing issues for which they were not originally designed, as is often the case (see Curet 2003, 2005; Davis 1988; Keegan 2004; Oliver 1999).

The application of Rouse's model for the archaeological record of Puerto Rico is particularly important because it was here that he articulated his original sequence, which was then projected onto the Greater and Lesser Antilles. Rouse's chrono-cultural framework is presented as a sequence of complexes or styles (normatively defined), articulated hierarchically into subseries and then into series, with finite geographical and temporal boundaries. According to Rouse's (1992a) latest scheme, the "marching" succession of cultural complexes of Puerto Rico starts around 1000 BC, when the first wave of migrants, grouped within the Corosan Ortoiroid subseries of the Archaic Age, reached the island (figure 2.1).[1] Around 400 BC, Cedrosan Saladoid peoples arrived in Puerto Rico and quickly eliminated, acculturated, or displaced its previous Archaic inhabitants westward. These peoples, whose earliest style of pottery was named Hacienda Grande, were presumably the ones that introduced ceramics and agriculture to Puerto Rico and the rest of the Antilles. The finding of a cultural manifestation in Vieques, named the "Huecoid" by Chanlatte Baik and Narganes (1980), led Rouse to postulate the presence of the Huecan Saladoid subseries. The Huecans were also supposedly eliminated or driven east by the Hacienda Grande people, who continued to operate in ethnic isolation during its latest phase, developing the Cuevas style of pottery around AD 400. This style lasted between AD 400 and 600, until the development of what Rouse termed the Ostionoid series, reflected by the diversification of the Cuevas in the eastern portion of the island into the Monserrate style of the Elenan Ostionoid subseries and, in the west, into the Pure Ostiones style of the Ostionan Ostionoid subseries. Around AD 900, the Monserrate changed into the Santa Elena style, and around that same date, the Pure Ostiones transformed into the Modified Ostiones. By AD 1200, both the Elenan and the Ostionan Ostionoid subseries developed into the Chican Ostionoid subseries, which has been associated with the rise of the Taíno culture on the island. The development of this subseries is reflected by the change of the Santa Elena style into the Esperanza style in eastern Puerto Rico and of the Modified Ostiones into the Capá style in the west.

The absolute chronology of this scheme was originally constructed using thirteen dates that Rouse (1963; Rouse and Alegría 1979) recovered from eight sites on the island, to which nine others were added later (Rouse and Allaire 1978). With the purpose of determining if the temporality assumed in the sequence established in Rouse's model holds up to the evidence recently generated in Puerto Rico, we conducted a survey of radiocarbon assays from the island that produced a corpus of more than five hundred dates spanning all of the major cultural components defined thus far. With the use of such a substantive database, we address in this chapter the problems of employing the pottery-based scales of temporality assumed in Rouse's model to illustrate how time, as an archaeological dimension, needs to be rethought in the construction of Antillean precolonial histories. To do this, we first discuss time and the problems associated with its compartmentalization. We then use that discussion as a platform with which to explore three major issues in the archaeology of Puerto Rico: the Pre-Arawak/Huecoid-Saladoid interface (Rodríguez Ramos 2005), the "La Hueca problem" (Oliver 1999), and the shifts in settlement patterns and the organization of communities registered during the Ceramic Age in the south-central part of the island (Torres 2005).

## Time Matters, Matters of Time

Time and space are two of the primary archaeological dimensions for understanding societies of the past. When we consider the nature of the archaeological record, especially on a regional level, we are challenged to model and explain the interrelationships and behavior of past societies in terms of social organization and development through time. A critical component of this endeavor involves the way in which we conceptualize time and control for it within particular modeling processes. Methodologically, archaeologists often manage and attempt to control for time by breaking it down into "meaningful" cultural frameworks in which it is synthesized and segmented into homogenous social, temporal, and spatial units (that is, a classic culture-historical approach). In the Caribbean, the construction of time has been primarily based on the spatio-temporal distribution of ceramic types (see Rouse 1992a) and social development measured by characteristics conceived to be inherently associated with these categories (for example, nucleated households and the construction of ballcourts and plazas). Admittedly, there is utility to some of this nomenclature, for it gives us a means to discuss the past in a generalized and comparative context.

However, this treatment of time is problematic in several ways. Because individuals within a society do not exist in a vacuum, social identities and relationships are built at varying scales of interaction and association within physical places through time (Bender 1993; Pauketat 2001). In this sense, culture and society can *both* be seen as continuous phenomena and always in the state of becoming or emerging. It is interesting that archaeologists often do not address or attempt to accommodate for time, space, and people in a fashion conducive to this perspective. One avenue that archaeologists have pursued to alleviate this issue is to utilize physical or absolute time (primarily through radiocarbon dating or other chronometric techniques) to help "control" for our reliance on artifact-based chronologies to substantiate our interpretations. Although critically important (and a solution to "typological" time), our ability to apply physical time is not without its own problems (Fitzpatrick 2006; Oliver 1999).

In terms of archaeological evidence and its interpretation, certain processes are inherent to particular socio-spatial (for example, households, villages, localities, regions, and so forth) and temporal scales (for example, decades, generations, and so forth) (Bailey 1983; see also Knapp 1992 for a neohistorical perspective). Archaeologically, these scales will represent varying degrees of resolution that require us to consider the potential temporal levels at which certain social processes operate. From this perspective, social change can occur in a gradualist or punctuated fashion and at a variety of spatial and temporal scales. In the case of the latter, these changes can potentially occur beyond the observable capabilities of the archaeological record in the Caribbean. However, by the same token, so-cial changes are not typically characterized by synchronic events, and the processes responsible for the archaeological record and the empirically observable data from which we can generate inferences about the past are created by diachronic processes (see Binford 1981).

Hence, time is process and operates in multiple dimensions and scales through social interactions related within and to the construction of space and place (Ingold 2000; Bender 1993). As our archaeological questions become increasingly specific (especially addressing issues related to cau-sality, the tempo and pace of social change, and other processes that oc-cur at relatively small spatio-temporal scales), firm chronological control also becomes increasingly important. For instance, in the case of studying regional interaction, there are many questions that can only be answered by examining groups of contemporaneous sites. However, there are other questions where the examination of social process(es) through time can-not be based on individual sites but requires reasonable samples of sites irrespective of contemporaneity (Plog and Hantman 1990). For example, Haviser's (1997) analysis of Early Ceramic Age settlement strategies, which shows a distribution of sites in the Caribbean through time based on ra-diocarbon dates, presents an example of examining relatively large-scale processes with relatively coarse temporal resolution. Likewise, site-based research necessitates tighter chronological controls in order to address site-specific questions (for example, Curet 2004; Garrow et al. 1995; Rob-inson et al. 1985). However, problems emerge when we move beyond the level of the site and attempt to address contemporaneity in cases in which temporal control is limited to a geographically patchy sampling of radio-carbon dates and "typological" time based on artifact-centered chronolo-gies that span, in most cases, over three hundred years.

Upon an examination of the utilization of "typological" time in Puerto Rico (and other parts of the Caribbean) in terms of ceramic-based peri-ods, we are faced with several issues. From a conceptual standpoint, the "typologization" of time homogenizes culture, collapses regional variation, and reduces social process to categories (Fabian 1983). Furthermore, the "typologization" of time assumes an essentialist and normative perspec-tive related to the distribution, organization, and development of social actors within the geographic and temporal realities in which they exist. Methodologically, this translates to the utilization of ceramic styles by ar-chaeologists that suggest that social/cultural groups begin and terminate at the spatio-temporal bounds of the typological categories. Congruently, it is often assumed that social changes are homogenous through time and space. Related to this use of time is the widespread notion that while there are some correlations between social phenomena and ceramic types (for example, the development of nuclear households, the construction of formalized ballcourts and plazas, and so forth), there is a temporally perfect and positive correlation between these elements. This is problematic, as transitional changes in ceramic styles may neither fall within the temporal categories assigned to them nor represent changes in social complexity based on an array of assumed related diagnostic traits that may or may not exist within this context. Critically, by utilizing ceramics, one is controlling for similarities in material culture that may or may not share processually, spatially, or temporally proximal relationships.

Additionally, there may be sites whose ceramic assemblages represent variations in style that may actually be contemporaneous and constitute regional or village variants, or the idiosyncratic products of individuals rather than chronological variation (Thomas and Ehrich 1969). Another issue may be the persistence of a particular style, indicative of a distinct period in a given area that continues far beyond its categorical reference points as the result of cultural vicissitudes of a particular community and may not be applicable over substantial geographical spaces (Thomas and Ehrich 1969). Furthermore, the use of pottery for defining cultures might introduce a gender bias, as this technology has been most commonly related to female activities. Finally, for establishing the different units based on pottery, decorated sherds have usually been employed, although these often constitute less than 10 percent of Caribbean collections (Keegan 2004: 39).

All of these time-related concerns are highly pervasive in the archaeology of Puerto Rico and the Antilles in general. To illustrate how such chronological issues are ingrained in our interpretations of the archaeological record of the island, we will now discuss briefly the construction of the radiocarbon database that served as the foundation for this work. We then present three cases that showcase different problems inherent in the use of "typological" time in Caribbean archaeology.

## The $^{14}$C Database from Puerto Rico

As previously noted, the present work is primarily based on a radiocarbon database from Puerto Rico, which is still in development. At present, we have been able to obtain around 560 assays from various locations on the island that were found in cultural resource management (CRM) reports, in academic publications, and through the generosity of colleagues who have provided unpublished results. Unfortunately, not every region is equally represented in the database, as there is a geographic skewness of dated sites toward those located from central to eastern Puerto Rico, especially in the coastal plains as opposed to interior mountains. In addition, the number of dates that have been collected from different sites is highly variable. For example, there are sites such as Paso del Indio that have more than forty dates, while others only have one or two dates. This shows that the data are biased at the site-specific and geographic levels, indicating the need to construct a more representative chronology for the island as a whole.

Another source of problems for interassemblage temporal comparisons with this database has been the dissimilar, and mostly ambiguous, ways in which pottery classes, types, or styles have been defined and the particular chrono-cultural indexes assumed in the different models developed in the island. For instance, Chanlatte Baik and Narganes (1980, 1983) do not consider styles as an important category in their model; they instead refer to analytical units that make reference to "cultures" (that is, *cultura Huecoide, cultura Saladoide*), while Rouse defined each culture by its style (Oliver 1999). This constitutes a major problem for comparison, as, for instance, Chanlatte Baik (1990) lumps what Rouse terms the Hacienda Grande and the Cuevas into a single category (that is, *cultura Saladoide/Agro II*), while Rouse considers each of those two styles to be distinct cultures that share a single path. In other cases, site assignments to specific series, subseries, or styles based largely on pottery are provided in the reports. But these associations are not based on detailed analyses, thus limiting the resolution and comparability of their insertions into particular chrono-cultural slots. The scale of resolution in the temporal placement of sites has also been variable, as some chrono-cultural designations are left at the level of series, while in others, the pottery styles have been recognized. Further problems for assigning sites to particular time periods or cultures are encountered when contexts that present mixed pottery styles are dated, which are quite commonly represented in the database. These are all issues that demand further attention, as they impose marked limitations on the levels of analyses that can be made with the available data.

The ways in which such dates have been collected and reported for different sites have also been highly variable, thus also limiting the comparability of the data (see Fitzpatrick 2006). However, owing to the high quantity of CRM projects conducted in the island, a significant number of radiocarbon assays have been recovered, which allows us to provide a more conservative analysis of the suite of dates that were used in this study. To increase the level of resolution of the database, we followed Spriggs's (1989) "chronometric hygiene" approach using a number of criteria applied by Liston (2005; see also Fitzpatrick 2006). We excluded from our analysis those assays that (1) lacked provenience information; (2) did not overlap at a 2σ range with the rest of the dates recovered from associated contexts within the sites; (3) were made from materials that tend to provide unreliable results, particularly those that that inhabit karstic haystack hills of the island (for example, land snails), because of problems with carbon uptake that can introduce age anomalies in such samples (Goodfriend and Hood 1983); (4) were not associated with materials defined at the style level; (5) were justifiably declared anomalous by its excavators; or (6) contained only one date from a site. This exercise diminished the amount of dates to around 360, all of which were calibrated (2σ) using the calibration datasets (IntCal 04 and Marine04) provided in the most recent version of the CALIB program (CALIB 5.0.1; Stuiver et al. 2005). No Delta-R correction was applied to marine shell samples during calibration.

## Time and the Pre-Arawak/Huecoid-Saladoid Interface

Time has been commonly overlooked when considering matters of culture contact and interaction, particularly when peoples that are ascribed to two distinct levels of social evolution are involved in the equation. This is particularly the case in the interface between the Pre-Arawak peoples of the island and the Huecoid and Saladoid newcomers. Due to the fact that the Pre-Arawak groups were supposedly hunter-gatherers organized in acephalous bands and the later Huecoid and Saladoid immigrants were tribal groups that practiced horticulture, it has commonly been assumed that because of the higher stage of social and technological evolution of the latter, they quickly displaced, eliminated, or acculturated with the former, giving substance to the horizontal boundary that separates them in Rouse's (1992a) scheme.

However, the short time span and the character of such relationships is put into question when considering what time has to say about the nature of Pre-Arawak societies and the temporal extent of their interactions with the later immigrants. According to Rouse's (1992a) most recent scheme,

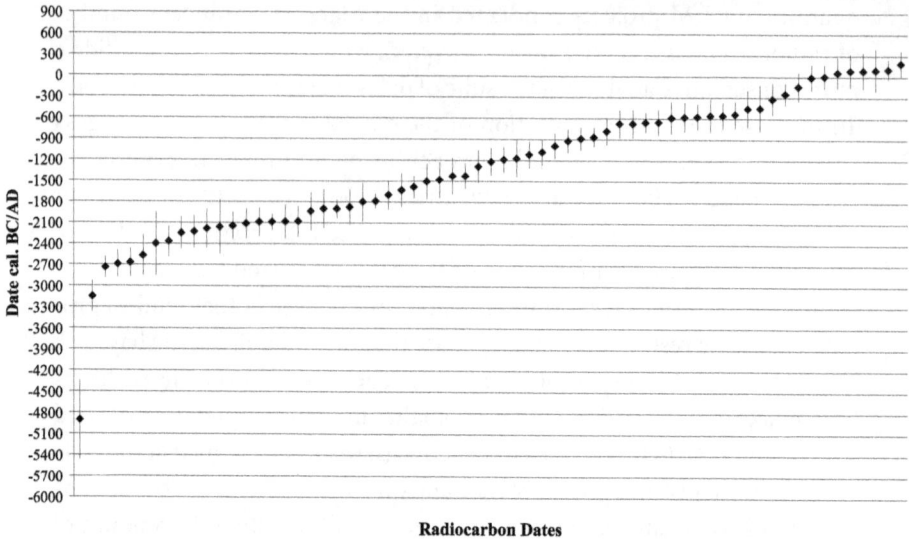

Figure 2.2. Spread of calibrated 2σ ranges of Pre-Arawak dates from Puerto Rico.

the earliest Pre-Arawak groups arrived to Puerto Rico from the Lesser Antilles around 1000 BC. However, the more than sixty dates that have recently been collected thus far from Pre-Arawak contexts present a continuous spread beginning around 2870–2680 cal BC (Beta-92891; Maruca site, Rodríguez López 2004), almost two millennia earlier than what was previously assumed (figure 2.2).[2] Radiocarbon assays from Pre-Arawak sites extend to cal AD 90–265 (Beta-8849; Yanuel 9 site, Tronolone et al. 1984), at least five hundred years later than what was established in Rouse's model.

Even though this chronometric data set shows that Pre-Arawak peoples were the ones who occupied Puerto Rico's territory the longest, it has commonly been assumed that they were static through time, experiencing little or no change over the millennia. This has been based both in their ascribed Archaic stage in Phillips and Willey's (1953) model of sociocultural evolution and on the supposed European encounter with hunting and gathering peoples in southwestern Haiti and western Cuba who arguably represented remnants of the groups displaced upon the arrival of Saladoid peoples to the islands (Rodríguez Ramos 2008). Both of these elements were used to suggest that time did not pass for the earliest inhabitants of Puerto Rico, who were argued to have remained socioculturally immutable after their arrival to the island (Rouse 1992a; Rouse and Alegría 1990). As previously noted, this staged entanglement of time and social evolution with the different "cultures" that inhabited the Antilles through time constitutes one of the most deeply rooted assumptions in Rouse's scheme.

Following this line of inquiry, the temporal placement of the origins of agriculture in the Antilles has been related to the entrance of the tribal South American immigrants to the islands who were sedentary and produced pottery, among other features that are supposed to be absent in hunting and gathering societies. However, the available temporal data shows that those first groups who arrived to Puerto Rico not only reached the island much earlier than previously thought but also seemed to have been practicing delayed-return economies and incorporating different cultigens throughout their occupational history. For instance, the Antillean botanical trinity of manioc, sweet potatoes, and maize, which was supposed to be introduced during the Saladoid, has all been documented during Pre-Arawak times in Puerto Rico at least since circa 1300 BC in the starch gain analysis conducted by Pagán Jiménez and colleagues (2005). Other products such as yellow sapote and sapodilla, probably obtained from Central America (Newsom 1993), have also been documented in later Pre-Arawak contexts (for example, Cueva María de la Cruz; Rouse and Alegría 1990). Evidence for human-induced forest fires, probably related to swidden agriculture, has been obtained from north-central Puerto Rico with dates that go back to 3490–3110 cal BC (Beta-54720; Laguna Tortuguero site, Burney et al. 1994). Other aspects such as pottery production and the articulation of village-type settlements also seem to predate the entrance of Saladoid and Huecoid peoples to the island (Rodríguez Ramos 2005, 2008). This demonstrates that there were transformative social processes affecting those Pre-Arawak groups, suggesting that the peoples encountered by Saladoid and Huecoid immigrants were much different from the ones that originally entered Puerto Rico. Thus, the sociocultural homogeneity of the different cultural manifestations that is portrayed within each of Rouse's squares blurs the high degree of vertical and horizontal variability that should have existed in the islands throughout their precolonial history.

As previously noted, Pre-Arawak groups were supposed to have disappeared quickly from the island upon the arrival of the South American invaders. The entrance of Saladoid groups to Puerto Rico has been dated at least as far back as 740–390 cal BC (I-13856; Tecla site, Narganes 2005), while the earliest calibrated date collected thus far for the Huecoid ranges

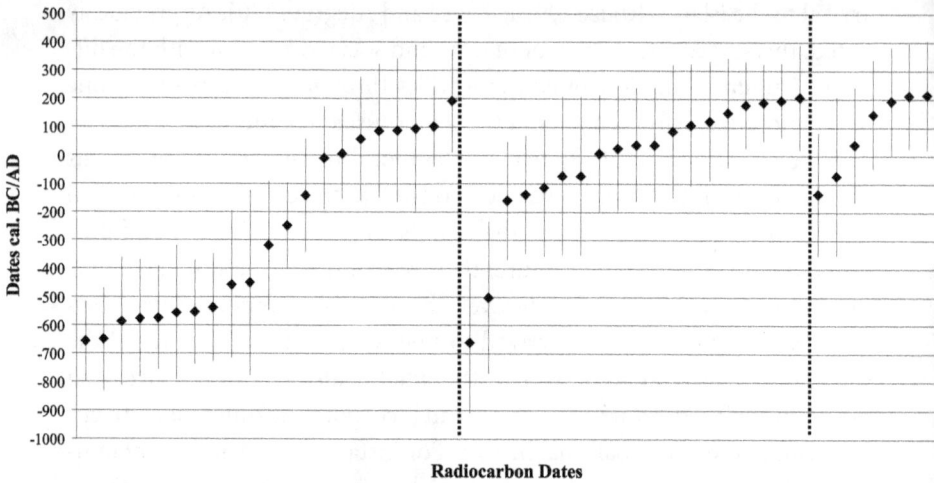

Figure 2.3. Calibrated dates (2σ) from late Pre-Arawak, early Hacienda Grande, and Huecoid contexts, showing their temporal overlap.

between 150 BC and AD 70 (I-14978; Punta Candelero site, Rodríguez López 1991a) (figure 2.3).[3] The fact that the latest dates for Pre-Arawak contexts in Puerto Rico extend up to circa AD 200 indicates that they coexisted on the island with these other groups for at least half a millennium. As such, this does not support their quick elimination, displacement, or acculturation as is assumed in Rouse's (1992a) model. Also noteworthy is the fact that some of the latest Pre-Arawak dates come from interior contexts, which might suggest a late inland incursion of these groups rather than a westward displacement as suggested by Rouse (1992a).

Interestingly, it is in the interior portion of the north-central section of the island where early pottery has been uncovered in association with Pre-Arawak assemblages, some of which present traits characteristic of Chican pottery and other Ostionoid styles (Rodríguez Ramos 2005; Rodríguez Ramos et al. 2008). Thus, it could be possible that some of those Pre-Arawak groups that moved inland continued to develop, eventually producing some of the pottery styles that have been lumped into the Ostionoid series (Rodríguez Ramos 2005). In fact, as seen in figure 2.4 (based on the 2σ spread of the available dates and on the continuity of the lithic production protocols and dietary practices that have been observed between the Pre-Arawak and the Ostionoid series), we can not falsify Chanlatte Baik's (1990) original hypothesis of the development of the Pre-Arawak into the

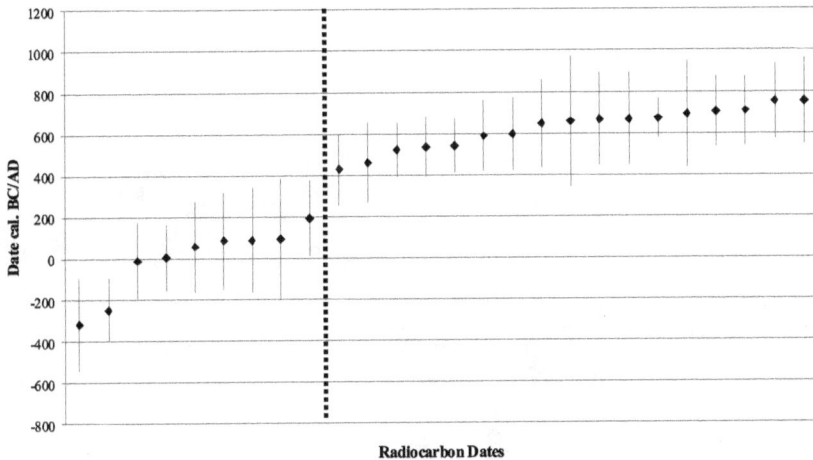

Figure 2.4. Temporal overlap at a 2σ level of late Pre-Arawak and Early Ostionoid dates.

different Ostionoid manifestations as a result of their interaction with both the Huecoid and the Saladoid peoples on the island.

All of this temporal evidence shows that when we disentangle time from developmental notions of culture and society, we are left free to examine the ways in which groups with different structural and/or cultural organizations might have interacted. The pervasive temporalization of cultural and/or social aggregates based on their supposed stage of development, and the preconceived biases that are involved when addressing the nature of their interactions on the basis of such premises, are aspects that limit our understanding about the myriad of possible scenarios that might have been possible in precolonial contact situations and interaction dynamics, such as the one previously discussed.

## Time and the "La Hueca Problem": A View from the La Hueca–Sorcé Site

The case of the La Hueca–Sorcé site on Vieques Island brings to the forefront the failures of the normative, culture-historical approach, particularly in its application for reconstructing cultures and histories based on intrasite scales of analyses, as was originally the case of what Oliver (1999) termed the "La Hueca problem." The La Hueca–Sorcé site contains perhaps

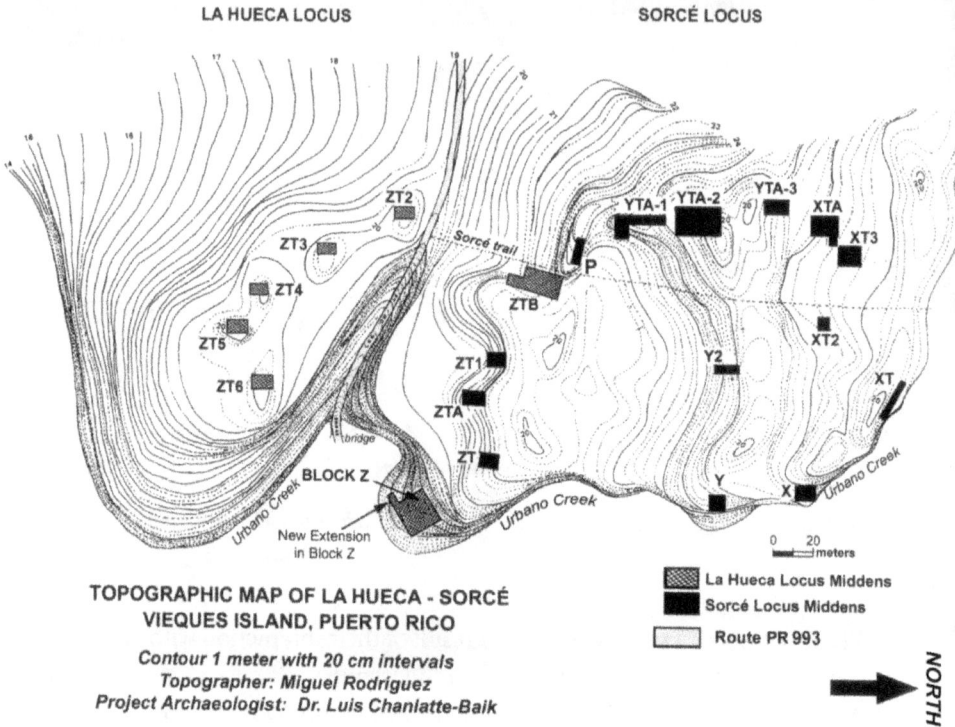

Figure 2.5. Map of the La Hueca–Sorcé site, showing excavation blocks (adapted from Chanlatte Baik and Narganes 1983).

the most controversial of all archaeological complexes in the Caribbean, in terms of both stylistic (cultural) identity and chronology. The evidence found here shows that there is either something very wrong with Rouse's (1992a) model or something amiss about the ninety-four radiocarbon dates from La Hueca–Sorcé. Since the key problems surrounding the definition and timing of the La Hueca complex have already been discussed by Oliver (1999) and Rodríguez Ramos (2001) at length elsewhere, these will not be reiterated here. Instead, we focus on the interpretation and implications of the large suite of radiocarbon dates thus far recovered from La Hueca–Sorcé.

To summarize, the La Hueca–Sorcé site has two major occupational loci, each with its own distinctive cultural complex (or style). These are completely segregated in space (figure 2.5), with La Hueca characterized by its distinctive ZIC (zone-incised crosshatch) ware, lithic reduction

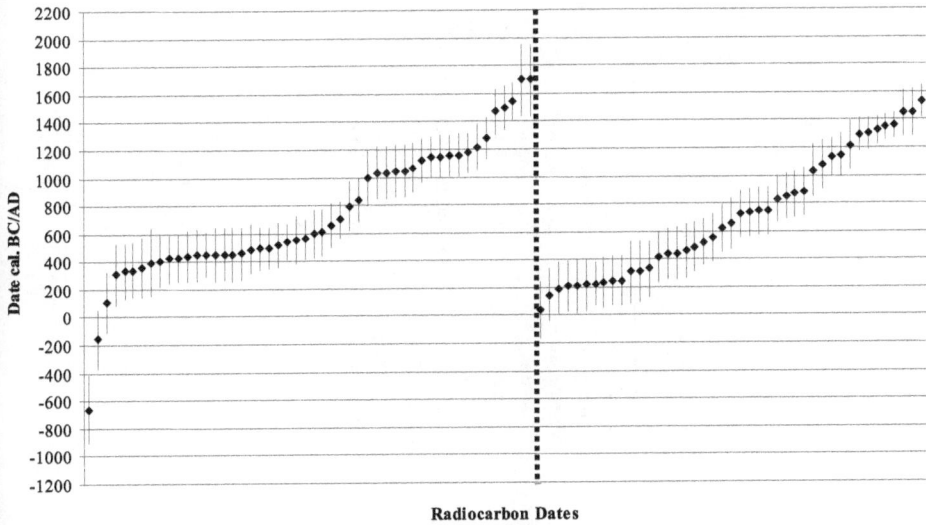

Figure 2.6. Temporal spread of dates from the La Hueca–Sorcé site, by locus.

sequences, and microlapidary production, whereas Sorcé is typified by its WOR (white on red) ware, its own lithic tool kit (far less varied than La Hueca), and a limited and stylistically distinct microlapidary work. It thus can be argued with some degree of confidence that indeed the artifact assemblage from La Hueca is best regarded as a separate Huecoid series, whereas Sorcé is best associated with the Saladoid series. In short, Chanlatte Baik's (1990) argument for a distinct (Agro I) Huecoid cultural manifestation is better supported than Rouse's (1992a) placement of La Hueca in a distinct subseries (Huecan) but deriving from the Saladoid series. The latter assumed that La Hueca and related styles (for example, Hope Estate, Punta Candelero, and so forth) diverged from a common Saladoid ancestry, whereas Chanlatte Baik and, more recently, others correctly argue that La Hueca has a different and separate developmental history (hence being what Rouse would call a separate series—Huecoid).

The situation described for the La Hueca and Sorcé loci is not, in itself, remarkable until one considers two simple observations. First, the forty-four radiocarbon dates obtained from the La Hueca (Huecoid) locus (table 2.1) excavations substantially overlap with the forty-nine assays recovered from the excavated units in the Sorcé (Saladoid) locus (table 2.2). Second, on both loci, the assays run though very long time spans, indicating that both occupations not only coexisted side by side but did so for more than a thousand years (figure 2.6).

Table 2.1. Huecan (Agro I) Dates, La Hueca Locus, Vieques

| Sample no. | Block/unit | Depth from surface of unit (cm) | Radiocarbon years BP | Material dated | Calibration curve (2σ) |
|---|---|---|---|---|---|
| I-11,142 | Block Z: Z-20 | 20–40 | 405±75 | Wood | AD 1410–1647 |
| I-10,549 | Block Z: Z-9 | 60–70 | 1,525±85 | Marine shell | AD 690–1035 |
| I-10,553 | Block Z: Z-9 | 150–160 | 1,565±80 | Marine shell | AD 680–1000 |
| I-11,321 | Block Z: Z-V | 160–170 | 1,845±80 | Charcoal | AD 135–470 |
| I-11,320 | Block Z: Z-W | 160–170 | 1,770±80 | Charcoal | AD 70–430 |
| I-11,141 | Block Z: Z-16 | 160–180 | 1,705±80 | Charcoal | AD 135–535 |
| I-11,322 | Block Z: Z-X | 170–180 | 1,945±80 | Charcoal | 160 BC–AD 240 |
| I-10,980 | Block Z: Z-11 | 190–200 | 1,735±85 | Charcoal | AD 85–530 |
| I-10,979 | Block Z: Z-8 | 200–210 | 1,820±85 | Charcoal | AD 20–405 |
| I-11,140 | Block Z: Z-15 | 200–220 | 1,730±80 | Charcoal | AD 90–530 |
| I-11,139 | Block Z: Z-15 | 240–260 | 1,800±80 | Charcoal | AD 55–410 |
| I-15,185 | New Extension, Block Z: C-12 | 60 | 540±80 | Charcoal | AD 1280–1620 |
| I-15,186 | New Extension, Block Z: C-10 | 80 | 520±80 | Charcoal | AD 1290–1620 |
| I-15,187 | New Extension, Block Z: B-9 | 100 | 690±80 | Charcoal | AD 1190–1420 |
| I-15,188 | New Extension, Block Z: A-9 | 150 | 700±80 | Charcoal | AD 1180–1410 |
| I-11,189 | New Extension, Block Z: B-9 | 160 | 790±85 | Charcoal | AD 1040–1385 |
| I-15,238 | New Extension, Block Z: B-10 | 190 | 570±80 | Charcoal | AD 1270–1455 |
| I-15,239 | New Extension, Block Z: B-10 | 200 | 660±80 | Charcoal | AD 1220–1420 |
| I-15,240 | New Extension, Block Z: B-10 | 210 | 630±80 | Charcoal | AD 1260–1440 |
| Beta-129,948 | Block Z (newest sample) | n/a[a] | 1,810±60 | Charcoal | AD 70–380 |

| I-18,449 | Block Z (newest sample) | n/a[a] | 1,740±80 | Marine shell | AD 470–820 |
|---|---|---|---|---|---|
| I-18,448 | Block Z (newest sample) | n/a[a] | 1,710±80 | Marine shell | AD 530–880 |
| I-18,661 | Block Z (newest sample) | n/a[a] | 1,670±80 | Marine shell | AD 570–905 |
| I-18,660 | Block Z (newest sample) | n/a[a] | 1,650±80 | Marine shell | AD 590–930 |
| I-18,450 | Block Z (newest sample) | n/a[a] | 1,640±80 | Marine shell | AD 610–940 |
| I-18,723 | Block Z (newest sample) | n/a[a] | 1,500±80 | Marine shell | AD 715–1050 |
| I-18,662 | Block Z (newest sample) | n/a[a] | 1,480±80 | Marine shell | AD 720–1070 |
| I-18,724 | Block Z (newest sample) | n/a[a] | 1,350±80 | Marine shell | AD 890–1240 |
| I-15,241 | Block Z (newest sample) | n/a[a] | 1,880±80 | Marine shell | AD 340–680 |
| I-12,856 | Block Z-T-B: C-8 | 80 | 1,810±80 | Charcoal | AD 30–405 |
| I-12,859 | Block Z-T-B: C-4 | 100 | 1,880±80 | Charcoal | 45 BC–AD 340 |
| I-12,858 | Block Z-T-B: B-3 | 100 | 1,820±80 | Charcoal | AD 30–400 |
| I-15,241 | Block Z-T-B: I-7 | 120 | 1,880±80 | Marine shell | cal AD 341–678 |
| I-12,860 | Block Z-T-B: C-1 | 120 | 1,780±80 | Charcoal | AD 70–420 |
| I-12,745 | Block Z-T-2: L-9 | 20–40 | 1,560±80 | Charcoal | AD 270–650 |
| I-12,743 | Block Z-T-2: L-8 | 20–40 | 950±80 | Charcoal | AD 900–1260 |
| I-13,426 | Block Z-T-2: K-7 | 20 | 1,810±80 | Marine shell | AD 410–740 |
| I-12,744 | Block Z-T-2: K-9 | 20–40 | 1,640±80 | Charcoal | AD 235–590 |
| I-12,746 | Block Z-T-2: LL-9 | 20–40 | 1,600±80 | Charcoal | AD 260–610 |
| I-12,742 | Block Z-T-2: K-7 | 20–40 | 900±80 | Charcoal | AD 1000–1270 |
| I-13,427 | Block Z-T-3: H-4 | 20 | 1,840±80 | Marine shell | AD 380–710 |
| I-13,428 | Block Z-T-4: E-5 | 20–40 | 1,930±80 | Marine shell | AD 275–640 |
| I-15,242 | Block Z-T-5: H-10 | 40 | 1,230±80 | Marine shell | AD 1020–1300 |
| I-13,429 | Block Z-T-6: G-5 | 20–40 | 1,640±80 | Marine shell | AD 610–940 |

*Sources:* Data adapted from Narganes 1991, 2005; Chanlatte Baik and Narganes 1983; Oliver 1999.

*Note:* Calibrated dates are rounded off to the nearest decade, except those with years ending in 5 (half decade). Dates were calibrated with CALIB Rev. 5.0. No Delta R was applied. No $^{12}C/^{13}C$ ratios were reported for these samples.

[a] No depths were provided for these radiocarbon samples.

Table 2.2. Saladoid (Agro II) Dates, Sorcé Locus, Vieques

| Sample no. | Block/area depth[a] | Radiocarbon years BP | Material dated | Calibration curve (2σ) |
|---|---|---|---|---|
| I-13,425 | Block Area Z-T-A | 2,110±80 | Charcoal | 370 BC–AD 50 |
| I-14,814 | Block Area Z-T-A | 1,240±80 | Marine shell | AD 1020–1300 |
| I-14,815 | Block Area Z-T | 1,380±80 | Marine shell | AD 840–1210 |
| I-14,816 | Block Area Z-T | 1,350±80 | Marine shell | AD 890–1240 |
| I-18,970 | Block Area Z-T-A (new) | 1,765±70 | Marine shell | AD 470–780 |
| Beta-129,949 | Block Area Z-T (new) | 1,920±60 | Marine shell | AD 340–630 |
| I-12,857 | Block Area Z-T-B | 1,580±80 | Charcoal | AD 260–640 |
| I-11,319 | Block Area YTA-1 | 1,915±80 | Charcoal | 110 BC–AD 320 |
| I-11,685 | Block YTA-1: L-36 | 1,740±75 | Charcoal | AD 90–530 |
| I-11,317 | Block YTA-1: L-5 | 1,615±75 | Charcoal | AD 260–600 |
| I-11,316 | Block YTA-1: G-5 | 1,555±75 | Charcoal | AD 350–645 |
| I-11,318 | Block Area YTA-1 | 1,490±75 | Charcoal | AD 420–660 |
| I-11,926 | Block Area YTA-2 | 1,720±80 | Charcoal | AD 130–530 |
| I-11,925 | Block Area YTA-2 | 1,665±80 | Charcoal | AD 140–570 |
| I-11,686 | Block Area YTA-2 | 1,575±80 | Charcoal | AD 260–640 |
| I-11,927 | Block Area YTA-2 | 1,565±80 | Charcoal | AD 265–645 |
| Beta-129,950 | Block YTA-2 (new) | 1,680±60 | Marine shell | AD 600–860 |
| I-11,687 | Block YTA-2: I-22 | 1,565±75 | Charcoal | AD 260–650 |
| I-10,547 | Block Area YTA-3 | 1,575±85 | Charcoal | AD 260–650 |
| I-10,551 | Block Area YTA-3 | 1,210±85 | Marine shell | AD 1025–1330 |
| I-10,552 | Block Area YTA-3 | 1,230±80 | Marine shell | AD 1020–1310 |
| I-16,176 | Block Area Z-T-B P (new) | 1,270±90 | Charcoal | AD 620–970 |
| I-16,153 | Block Area P (new) | 2,590±90 | Charcoal | 910–410 BC |
| I-15,241 | Block Area P (new) | 1,880±80 | Marine shell | AD 340–680 |

| | | | |
|---|---|---|---|
| I-16,151 | Block Area P (new) | 1,700±80 | Charcoal | AD 140–535 |
| I-16,152 | Block Area P (new) | 1,650±80 | Charcoal | AD 220–590 |
| I-16,154 | Block Area P (new) | 1,620±80 | Charcoal | AD 250–600 |
| I-16,174 | Block Area P (new) | 1,600±80 | Charcoal | AD 260–610 |
| I-16,173 | Block Area P (new) | 1,590±80 | Charcoal | AD 260–630 |
| I-16,175 | Block Area P (new) | 1,450±80 | Charcoal | AD 420–765 |
| I-18,973 | Blck Area X-T-3 (new) | 1,960±110 | Marine shell | AD 180–660 |
| I-18,726 | Block Area X-T-3 (new) | 1,810±80 | Marine shell | AD 410–740 |
| I-18,972 | Block Area X-T-3 (new) | 1,715±70 | Marine shell | AD 540–840 |
| I-18,725 | Block Area X-T-3 (new) | 780±80 | Marine shell | AD 1420–1680 |
| I-14,850 | Block Area X-T-3 | 1,340±80 | Marine shell | AD 990–1240 |
| I-14,847 | Block Area X-T-3 | 1,220±80 | Marine shell | AD 1030–1310 |
| I-14,848 | Block Area X-T-3 | 1,190±80 | Marine shell | AD 1040–1330 |
| I-14,813 | Block Area X-T-3 | 1,180±80 | Charcoal | AD 670–990 |
| I-14,846 | Block Area X-T-3 | 1,150±80 | Marine shell | AD 1070–1390 |
| I-14,845 | Block Area X-T-3 | 1,080±80 | Marine shell | AD 1165–1440 |
| I-10,550 | Block Area X | 1,505±85 | Charcoal | AD 360–670 |
| I-10,548 | Block Area X | 1,440±85 | Charcoal | AD 420–770 |
| I-15,727 | Block X (new) | 1,350±80 | Marine shell | AD 890–1240 |
| I-15,728 | Block X (new) | 1,340±80 | Marine shell | AD 890–1240 |
| I-15,719 | Block X (new) | 1,320±80 | Marine shell | AD 920–1255 |
| I-15,718 | Block X (new) | 1,270±80 | Marine shell | AD 990–1290 |
| I-15,658 | Block X (new) | 470±80 | Charcoal | AD 1300–1635 |
| I-15,657 | Block X (new) | 410±80 | Charcoal | AD 1330–1650 |
| I-15,656 | Block X (new) | 300±80 | Charcoal | AD 1430–1952[b] |

*Sources:* Data adapted from Narganes 1991, 2005; Chanlatte Baik and Narganes 1983; Oliver 1999.

*Note:* Dates are rounded off to the nearest decade, except those with years ending in 5 (half decade). Dates were calibrated with CALIB Rev. 5.0. No Delta R was applied. No depths and $^{12}C/^{13}C$ ratios were reported for these samples.

[a] Dates not yet published.

[b] Modern sample.

However, these observations remain tentative, as not a single date from this site has been contextualized stratigraphically. As Jack Liu (in the radiocarbon laboratory at the Illinois State Geological Survey) once remarked to Oliver back in 1989, all dates are "correct" within the limits of calibration and uncertainties. "Wrong" dates can always be explained (for example, mixing of carbon of different ages, contamination, stratigraphic displacement, and so forth), but the date itself does not "lie." The lack of precise stratigraphic placement, whether the stratum is "natural" or "anthropogenic," is undoubtedly what makes any serious attempt to explain these anomalous assays a dubious exercise. Nevertheless, we will risk providing some comments and observations on a few of the apparent patterns exhibited by each data set (tables 2.1 and 2.2).

Excluding a clearly intrusive, near-surface sample from a wood log (I-11,142), the earliest date for the La Hueca locus is cal BC 160–AD 240, and the latest is cal AD 1290–1620 (table 2.1). At face value, the occupation at the La Hueca locus lasted a minimum of 1,450 or a maximum of 1,860 years. For the Sorcé locus, the spread of calibrated dates is even longer. Excluding one modern sample (I-15,656), the dates range from cal 910–410 BC to cal AD 1420–1680 (table 2.2). This spread represents a minimum of 2,090 to a maximum of 2,330 years of continuous occupation.

Looking at the La Hueca locus, Block Z (a sloped midden deposit), Oliver (1999) noted that the set of *charcoal* dates coming from units Z-8, Z-11, Z-15, Z-16, Z-V, Z-W, and Z-X (table 2.1) ranged between cal 160 BC–AD 135 to no later than cal AD 240–535. The depth, taken from the surface of the north corner (=0 cm) of each 4 × 4 meter unit, allowed Oliver (1999) to roughly guess that all proceeded from below a thin, sterile sand stratum described by Chanlatte Baik, possibly the result of slope-wash erosion created by a tropical storm event. However, above the sterile sand layer, the archaeological materials are regarded by Chanlatte Baik (1990) to be essentially the same style as those below (which were radiocarbon dated). The upper deposition, following the erosion event, should be no earlier than around cal AD 400. One of two marine shell dates (table 2.1: I-10,549) that can be placed above the erosion event is consistent by dating to cal AD 690–1035, whereas the other (I-10,553) comes from a unit and depth that would place it below the sand facies. That date (cal AD 680–1000) is not consistent with the suite of dates noted above.

After 1983, Chanlatte Baik and Narganes added an extension to Block Z (table 2.1, New Extension, Block Z; see figure 2.5), which is located toward the base of the slope. Remarkably, all the dates, regardless of depth, returned a narrower and different spread of calibrated dates between cal AD 1180 and 1440, two of which fall around AD 1620 in the upper 2σ range. For Unit B10, the three dates between 190 and 200 centimeters below unit surface (I-15,238 to I-15,240) are essentially the same: AD 1220–1460. These later dates, while internally consistent in this new extension of Block Z, cannot be easily reconciled with the suite of earlier dates obtained from the units excavated higher on the midden slope (and below the sand stratum). The two marine shell dates above the sterile sand layer are earlier (AD 680–1035), but there is a small chance that they overlap (at 2σ, that is, around AD 1000–1040). Nevertheless, the new extension to Block Z yielded what Chanlatte Baik and Narganes (1983) called "typical La Hueca artifacts."

The newest dates reported for Block Z by Narganes (Narganes 2005; see table 2.1, "newest samples"), with two exceptions (Beta-122,948 and I-15,241) present a pattern that for the most part fills the gap between AD 530 and AD 1000. It is not known (to Oliver) which excavation units these dates refer to, or the depth or whether these are from the new extension to Block Z. What is interesting is that all the newly reported dates were from marine shells, while one of the two exceptions noted (Beta-122,948) was from charcoal. Presumably, all of these dates are still associated with La Hueca–style components or assemblages. The dates from these latter sets, particularly those postdating the sterile sand and those from all levels in the new extension from Block Z, were ignored by Rouse (1992a) in his last revision of the cultural chronological chart. Aside from the lack of stratigraphic and contextual details, there are still far too many dates (in Block Z alone) and from different units to suggest that such a large number of them ought to be casually dismissed as intrusive.

What about the other La Hueca dated middens? The Z-T-B excavation (table 2.1) yielded five dates that were all consistent with the range of dates obtained in Block Z below the sterile sand layer, and all came from depths between 80 and 120 centimeters below the surface. The Z-T-2 excavation is more erratic given that all samples came from a depth of 20–40 centimeters. Blocks Z-T-3, Z-T-4, Z-T-5, and Z-T-6 have only one date each, with the first two having dates that correlate with Block Z dates below the sterile sand layer and with those of Z-T-5 and Z-T-6 matching some of the (later) dates of Z-T-2 and the "newest samples" from Block Z.

The situation for Sorcé will not be discussed in detail. For now, one date seems to be an outlier in relation to the spread of the earliest dates from La Hueca: cal 910–410 BC (figure 2.1: Block P). However, it should be noted that this date is contemporaneous with others recovered from Trants in Montserrat and Hope Estate in Saint Martin, indicating the need to further examine its veracity. Other than this date, the pattern is closely comparable to that of La Hueca. But again, the lack of specific provenance details and contextual associations makes it hard to make a further assessment, other than to say that Saladoid materials ranging from Hacienda Grande and Cuevas (or Agro II) to, perhaps, early Elenan (Monserrate style), are linked to this rather large temporal spread.

It is clear that to evaluate the ninety-four radiocarbon dates from La Hueca–Sorcé, far more contextual data is required, and details about site formation and "deformation" processes need to be elucidated (and published). It would also be useful to know how the charcoal samples were selected. For example, the seemingly erratic dates from Z-T-2 excavations could be the result of combining charcoal fragments with different ages from within each thick (20 centimeters) arbitrary layer. While some dates could be demonstrated to be unrelated to the age of the assemblages (or styles), it is just as likely at the moment that the Huecoid tradition (rather than "style") at La Hueca persisted (with limited changes) well after the temporal range arbitrarily imposed in Rouse's (1992a, 1992b) chronological charts. Much of the same could be said about a longer persistence of the local Saladoid tradition at Sorcé and that the bearers of both traditions did live side by side with little mutual influence in their material culture, that is, retaining their particular identities. Such statements, as hypotheses, could only be raised if we dare to break the shackles imposed by the rigid and mechanistic way in which, until recently (and perhaps still now), archaeologists "lodge" their cultural and social units of analyses (styles, societies, and so forth) into preconstructed time-space niches. The case of La Hueca–Sorcé is a good example, at a site level, of what we are beginning to see at a larger, regional level in the Caribbean.

## Time and Space: A View from South-Central Puerto Rico

As indicated in the previous example, when questions become more specific, the problems of "typological" time become increasingly evident. One area of archaeological inquiry that exemplifies this is the realm of settlement studies. On Puerto Rico, such studies have also relied on Rouse's

(1992a) chrono-cultural framework as a control for time, using ceramic styles registered for sites within a given region (Lundberg 1985; Maíz 1984; Rodríguez López 1984, 1990; Torres 2001, 2006). Problematically, stylistic information is unavailable for many sites, thereby decreasing their temporal resolution. This is, in part, a product of variability in the recording and sampling strategies of individual researchers working in the region and when the information was collected. Hence, in many cases, settlements can only be represented temporally at the level of the subseries or series. While this scale is useful for providing information regarding long-term social processes and the *potential* for intersettlement relationships within a given period, it cannot definitively support fine-scaled propositions regarding contemporaneity.

To elucidate some of the problems with settlement studies and the use of "typological" time to control for contemporaneity, let us consider sites from the south-central portion of Puerto Rico (figure 2.7). In this area, which includes the Portuguese, Cerrillos, and Canas river valleys, there are currently twenty-eight sites that possess definitive information regarding ceramic characteristics out of at least thirty-five or so recorded sites in the region (compare figure 2.7A with 2.7B through 2.7D). However, out of those, very few ceramic assemblages are identified to the level of style; most are left at the subseries level. This indicates that the temporal resolution of those sites identified to the subseries level ranges between three hundred years (Chican Ostionoid) and more than one thousand years (Cedrosan Saladoid).

Therefore, although the segmentation of time based on ceramics is useful for tracing the construction of cultural identities related to place and the potential for sociocultural relationships through time, it is problematic in many regards for fine-scale temporal control. For instance, based solely on pottery, we have no way of knowing if two closely related sites with similar ceramic assemblages reflect separate contemporaneous villages or if one represents the movement of people from one place to another (figure 2.7). Moreover, it is difficult to determine if the ceramic component at a given site represents the beginning or end of a particular period, causing a skewed perspective of the temporal affinity of a site within the contexts of social processes that are usually associated with transitions in material culture, as mentioned earlier. Critically, sites are often treated as being occupied for the entirety of a given period even though they may have been occupied for only a portion of that time. This is particularly evident

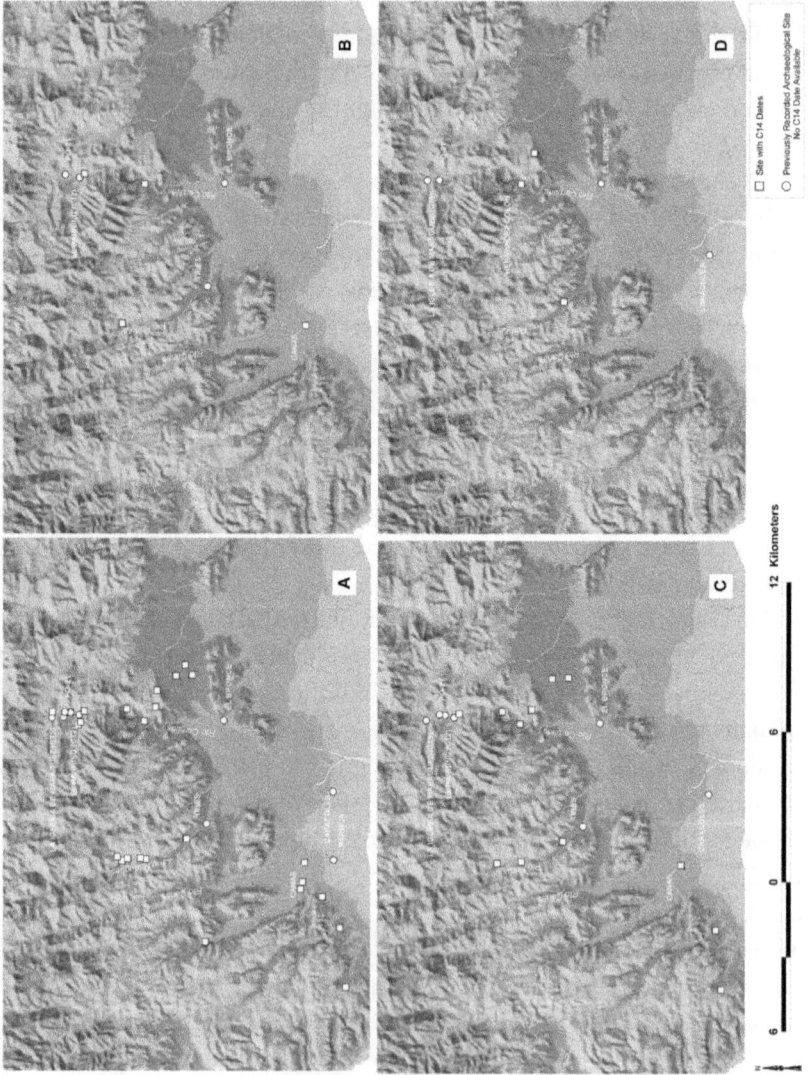

Figure 2.7. Sites in the Ponce area divided according to Rouse's periods: A, All sites; B, Period II sites; C, Period III sites; D, Period IV sites. The Archaic period is not shown, as only one site represents this period in the area.

in cases where ceramics from two chronologically contiguous periods are located in a single site. In these cases, sites are often treated as having been occupied for the entirety of both periods rather than at the end of one and the beginning of another (for an extended discussion see Plog and Hantman 1990). The problems associated with the use of "typological" time in the examination of multiple sites within a given region or locality can result in "map overestimation," in which sites that possess diagnostic traits of a given category are viewed as contemporaneous. This results in a potential overrepresentation of inhabited sites in an area at any one point in time (Ammerman 1981; Plog and Hantman 1990). This becomes a major methodological problem even in situations where the temporal categories are represented by relatively short periods of time (Plog and Hantman 1990) and where examining social interaction at relatively fine scale is a primary research question.

These factors become immediately apparent on a regional scale in south-central Puerto Rico when comparing radiocarbon dates against socio-temporal categories in the contexts of settlement studies. First, there is an overlap of time ranges for several of the sites based on the radiocarbon dates that suggests a potential for contemporaneity with those ranges and thus for social interactions at different points in time (figure 2.8). Critically, comparison of the date ranges and areas of overlap can be utilized

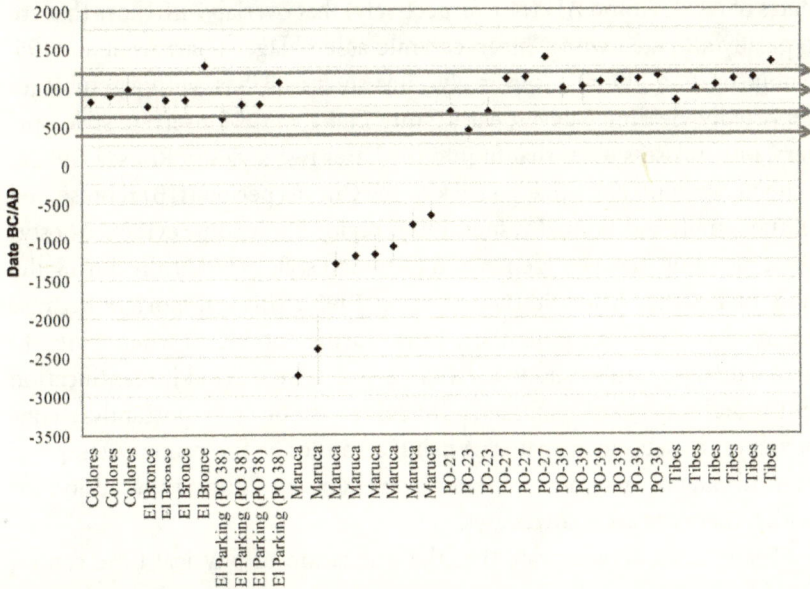

Figure 2.8. Radiocarbon dates for sites in the Ponce area.

to quantify this potential based on the extent of the overlap between sites. However, the concomitant habitation of some of those sites would not be apparent using "typologized" time because some of them presented different pottery styles. For example, the earliest dates for Hernández Colón (cal AD 409–863 [Beta-23902]; Maíz 2002) and PO-27 (cal AD 445–890 [Beta-23283]; Krause 1989) sites indicate that these have a high potential of being contemporaneous at least for some time, although both belong not only to two different styles (the former Hacienda Grande and the latter Pure Ostiones) but also to two different series (Saladoid and Ostionoid, respectively). Therefore, what becomes apparent through an examination of these radiocarbon dates is the continuous nature of social processes vis-à-vis human action independent of "typological" time frames. This observation lends strength to the notion of time and social process as continuous phenomena and not neatly compartmentalized or essentialized categories.

Further examination of the archaeological sites in the Ponce area with radiocarbon dates shows other interesting disparities in relation to Rouse's (1992a) socio-temporal model. In particular, it is clear that there are some potential discrepancies between reported ceramic styles and their temporal affiliation based on the radiocarbon dates. For instance, in regard to the site data for PO-39 and El Bronce, Elenan and Ostionan ceramics have been found in association with radiocarbon dates (mean radiocarbon dates of AD 1370 and AD 1140, respectively) that overlap with those that are typically affiliated with Chican ceramic assemblages (Garrow et al. 1995; Robinson et al. 1985). Conversely, sites such as Tibes, with assemblages reportedly yielding Cuevas, early Santa Elena, and Monserrate style pottery, only possess dates that fit into the latter two styles in Rouse's (1992a) scheme, with no dates that go back to the Cuevas period (IIB) represented in the sample. Additionally, sites in this region show some evidence of stylistic blending between Elenan and Ostionan styles specifically during the later half of the Ostionoid (Garrow et al. 1995). This is important because regional variants can potentially cause further problems when trying to associate a particular site to a given period or subperiod if identification of the archaeological material is difficult to establish. Importantly, a construction of settlement patterns using ceramics as the primary temporal control may or may not accurately represent sites during the period for which there are associated dates.

These cases demonstrate that the utilization of physical time can (1) help in the control of the material correlates related to particular social groups in space and time; (2) show that social process cannot be compartmentalized in neat spatial and socio-temporal packages; and (3) facilitate the integration of people in place and time. However, despite the obvious benefits of the use of absolute dating techniques in the realm of regional studies, there are still problems associated with their use. First and foremost is the issue associated with the small numbers of samples relative to the number of sites throughout the region. This is often a product of the level of investigation at the site and the availability of the data if samples do indeed exist. For instance, in this discussion only nine sites out of at least thirty-five for the Ponce area have radiocarbon dates, even though this is one of the areas with the highest representation of dated contexts in all of Puerto Rico. Further, the radiocarbon date ranges at 2σ (which extend to approximately three-hundred-year ranges in many cases), do not offer much better resolution than the typological designations at the subperiod level (Keegan 2004).

What the radiocarbon dates do offer, however, is a temporal range for human action within specific places that can be used to associate material correlates rather than the other way around. It will be through this interactive process, which utilizes absolute dating techniques to hone the relative dating of stylistic ceramic characteristics within specific geographical regions, that the study of settlement systems in the Caribbean will be able to address more fine-scaled social processes.

## Concluding Remarks

As has been made evident in the preceding analysis, the mistreatment of the time dimension in the archaeology of the Antilles has had significant repercussions in the ways in which we have interpreted the types of interactions, cultural relationships, and settlement patterns, among other aspects, of the precolonial peoples that inhabited the island. Each of these cases has demonstrated the need to work outside the boxes of Rouse's (1992a) model but also indicates the special attention that has to be paid to the temporal scale in which the particular phenomena under study should be addressed. Furthermore, we have seen that there is substantial horizontal and vertical variability within the boxes that encase the different cultural manifestations defined for the island and that such variability is usually eclipsed when sites are lumped in the pottery-based chrono-cultural units.

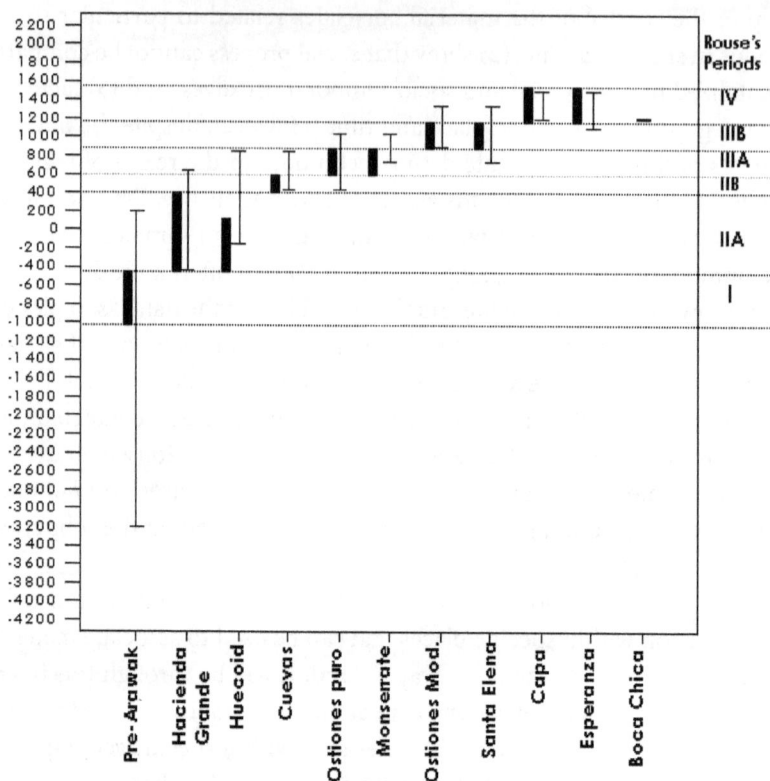

Figure 2.9. High-low logs showing the temporal distribution of the different manifestations in Puerto Rico, based on the midpoints of the 2σ spreads of the radiocarbon dates for each of them. The black boxes show the spread of such manifestations according to Rouse's (1992a) most recent scheme.

The database that forms the backbone on which our interpretations have been based has demonstrated the need to conduct similar surveys in the different islands (see Davis 1988 for an early example), as it is very likely that the recently generated data will make evident the inadequacy of reproducing a framework that was first developed more than half a century ago, experiencing little or no change since its original construction. The problems inherent in Rouse's (1992a) chrono-cultural scheme are particularly evident in Puerto Rico, where the available dates indicate that the absolute chronology of its cultural sequence is inaccurate. In order to demonstrate these errors, we took the midpoints of the 2σ spreads of the radiocarbon dates and produced box plots, which we then collapsed into high-low logs for each of the pottery styles to get a better panorama of their temporal distribution (figure 2.9). For comparison, we also include black boxes for the dates of each pottery style as established in Rouse's (1992a) most recent scheme.

At first glance, figure 2.8 shows that when we evaluate these newly generated dates for the island, many of the temporal extensions and phylogenetic relationships of the cultural manifestations defined by Rouse need to be reconsidered. According to Rouse, Hacienda Grande ceramics were manufactured between 400 BC and AD 400. However, as previously noted, if a recent date collected from Sorcé in Vieques is considered, there seems to be the possibility that this manifestation might have been even earlier, perhaps dating back to at least 600 BC. If this date is confirmed, it would be one of the earliest assays for a Cedrosan context in the Antilles and would support the hypothesis that native peoples moved northward into Puerto Rico first from a yet unknown location in South America (Fitzpatrick 2006; Keegan 2004; Rodríguez Ramos 2001, 2006). Based on the available evidence, the Hacienda Grande pottery seems to have been produced until at least circa AD 650, 250 years later than its previously estimated terminal date.

According to Rouse, the Huecoid manifestation was supposed to be in existence between 400 BC and AD 250. However, the evidence we have thus far seems to indicate that it lasted between 150 BC and at least AD 850, 600 years later than its supposed date of elimination or eastward displacement. Moving later in time, the Cuevas style of pottery was supposed to be produced between AD 400 and 600. However, the dates suggest that this style of Cedrosan pottery lasted between AD 450 and 850, 250 years later than the previous estimation for its terminal date. Interestingly enough, not only do the dates of the Cuevas span the full extent of the Huecoid, but its beginning seems to coincide temporally with that of the Pure Ostiones style, which supposedly marks the beginning of the Ostionoid series in the island.

The finding of these early dates for pottery associated with the Ostionoid series thus challenges the notion that it actually developed from the Cuevas, at least in western Puerto Rico. However, when considered regionally, the finding of early Ostionan ceramics in the occidental portion of the island should not be surprising, based primarily on the recovery of similar pottery in the southeastern Dominican Republic in sites such as La Iglesia, dating to around AD 350 (Ortega et al. 2004). Therefore, it could be possible that the Ostionan development reflects not a strictly intraisland evolution in western Puerto Rico of the Cuevas, as was suggested by Rouse (1992a), but rather the result of the interaction between societies in what Rouse (1992a) called the Mona Passage area. If this is the case, then the articulation of this passage area also predates its previously estimated origin, which according to Rouse was presumed to coincide temporally with the rise of the Ostionoid series. Pure Ostiones pottery is found on the island until at least AD 1050; this is 150 years later than its previous estimation.

Another important issue raised by the evaluation of [14]C dates in Puerto Rico has been the relative genetic relationship between the Monserrate and the Santa Elena styles. When we look at the available dates, we see that the start of both styles seems to be contemporaneous, beginning somewhere around AD 750. It is also interesting that the earliest dates for the Santa Elena pottery, which was supposed to be a style that arose in eastern Puerto Rico, actually come from sites in the south-central portion of the island, also casting doubts about the geographic origin of this pottery manifestation as was originally noted by Lundberg (1985). Rouse's (1992a) original temporal estimations for the latest pottery styles of the island, the Esperanza and the Capá, seem to be the ones more in concordance with the recently generated dates, beginning around AD 1100 and 1200, respectively. However, one issue that becomes evident when looking at the temporality observed in sites where both styles are mixed, versus those that present either style in isolation, is their seemingly rapid spread across the island. Unfortunately, at this point the directionality of such spread is not clear since the earliest dates for both styles come from contexts where both are present (which are mostly located in the north-central portion of Puerto Rico). Interestingly enough, this is the area where most evidence for early pottery in association with Pre-Arawak assemblages has been uncovered (Rodríguez Ramos et al. 2008). This again might indicate that some of the later cultural complexes identified in the island could represent the remains of developed Pre-Arawak peoples.

In addition to the temporal incongruities observed for each of the ceramic manifestations defined for Puerto Rico, one thing that becomes evident when looking at these date ranges is the coexistence of various pottery styles in the different periods established by Rouse (1992a), some of which contain up to five ceramic complexes. This high degree of horizontal variability casts doubts on the required cultural homogeneity assumed for each of those periods in Rouse's (1992a) model. This figure also shows that the different pottery styles defined for the island have varying levels of resolution, ranging from about three hundred years for the Monserrate to approximately one thousand years for the Hacienda Grande. This again makes clear that not all ceramic complexes should be used as relative temporal indices for addressing the same types of issues, some of which occur at even finer scales than the styles, as previously noted (see Curet et al. 2004; Oliver 2006: 477). Furthermore, the new dates generated for the island also seem to indicate that changes in pottery styles did not necessarily coincide with other social shifts, thus again showing the need to disentangle culture and society when addressing the temporality involved in such changes and their distribution.

All of this temporal evidence from Puerto Rico shows that in order to formulate models related to social development in the Caribbean and to elucidate the processes responsible for these developments at varying scales, there is a dire need for tighter chronological control and scrutiny of the conceptual frameworks in which we apply ceramic-based chronologies as culturally, spatially, and temporally homogenous and bounded units. However, we recognize the need to develop a system for establishing regional chronologies and measuring change at multiscalar temporalities. It is necessary to utilize physical time and associations in material culture to assess contemporaneity and control for artifact-reliant chronologies— an endeavor well under way. Yet there are no absolutes in chronological reconstruction, and we, as archaeologists, must not lose sight of the forest for the trees in a preoccupation with increasing chronological control by searching for the smallest divisible temporal unit: rather, it is the one that is best adapted to the nature of events and the scale at which we wish to investigate social process and the data available at hand (Bloch 1953; Plog and Hantman 1990). These must be utilized as generalized reference points and understood within the continuum of physical (or real) time that suggests that social processes are operating at both spatial and temporal scales outside of our "typological" constructions. Therefore, they can be used as a *general guide* to discuss coarse-resolution observations based on broad-based trends in social cultural traits but at the same time independent of them. Through this, it is possible to begin to discuss sociocultural developments within time and place, while developing more temporally controlled models (through dating techniques) of social development within discrete geographical contexts. Only then will we be able to understand the social and cultural reasons for the horizontal and vertical variability observed in

Puerto Rico and, very likely, in the rest of the insular Caribbean through time.

In sum, with this work we hope to have demonstrated that the "typological" form of time embedded in the normative perspective that has dominated Caribbean archaeology should be dramatically reconsidered, as it is evident that not all groups underwent the same social, cultural, or historical trajectories even in cases where they produced the same type of pottery. It is now time to determine whether we will continue to base our understanding of the pre-Columbian landscape of the Antilles using pottery as our primary guideline, or whether we will begin to open our eyes to more adequate and efficient ways of reconstructing the histories and cultural relationships of those people that constitute the foundation of the culture-historical stratigraphy of the Caribbean islands.

## Acknowledgments

We want to express our deepest gratitude to those archaeologists who produced the corpus of radiocarbon dates that served as the foundation of this paper, many of whom provided us with unpublished results of their work. We particularly want to thank those who personally sent us their dates, including Miguel Rodríguez, Luis Chanlatte Baik, Ivonne Narganes, Jaime Vélez, Juan González, Juan Rivera Fontán, Chris Goodwin, Sue Sanders, Daniel Koski-Karell, Tim Sara, Carlos Pérez, Peter Siegel, Marisol Meléndez, Hernán Ortiz, L. Antonio Curet, Edgar Maíz, and Marlene Ramos. Thanks also go to the personnel of the Puerto Rico State Historic Preservation Office (PRSHPO), especially to Mickey Bonini, and to Belford Matías and Lourdes Carrasquillo of the Consejo para la Protección del Patrimonio Arqueológico Terrestre de Puerto Rico. We also want to thank Scott Fitzpatrick and Ann Ross for their invitation to participate in the SAA symposium that led to this publication. This material is based upon work supported under a National Science Foundation Graduate Research Fellowship.

## Notes

1. The term "Pre-Arawak" was coined by Rodríguez Ramos (2007, 2008; Rodríguez Ramos et al. 2008) to make reference to what has traditionally been termed the "Archaic Tradition of Puerto Rico" (Alegría et al. 1955), in order to avoid the social and cultural implications that the homonym "Archaic" has imposed on their interpretation. However,

we recognize that the use of this concept is somewhat problematical due to its basis in linguistic designations, which are themselves a matter of contention. Hopefully, we can come up with a better term that addresses the horizontal and vertical variability of the original discoverers of the island in the near future.

2. Two earlier dates (one of 5200–4550 cal BC [Beta-29778] and other one of 3340–3030 cal BC [Beta sample number unknown]; Ayes 1993) were recovered from Angostura and the Hato Nuevo sites, respectively, both located in the Cibuco River valley. However, these are not included in the discussion because the first has an exceedingly large standard deviation in its radiocarbon age (±250) and does not overlap at a 2σ level with the rest of the Pre-Arawak dates, while the latter was obtained from a land snail (*Caracollus caracolla*). Both of these dates require further corroboration.

3. An earlier date was recovered from one of the Saladoid deposits ("P") of the La Hueca site in Vieques, dating between 910 and 420 cal BC (I-16153; Narganes 2005). Although the latest portion of its 2σ range coincides with the earlier part of the range of the earliest date for Tecla and is contemporaneous with other early assays obtained from sites such as Hope Estate and Trants, we still feel that it is necessary that such an early date be corroborated in Puerto Rico because of its singularity within this site.

# 3

# The History of Amerindian Mitochondrial DNA Lineages in Puerto Rico

JUAN C. MARTÍNEZ-CRUZADO

## Abstract

As the mode of inheritance of the mitochondrial DNA (mtDNA) is strictly maternal and nonrecombinatorial, the fast evolutionary rate of its control region allows a rapid accumulation of mutations that is useful for studies on the migratory and demographic events that gave rise to human populations across the world. In this chapter, I analyze 1147 bp of control and adjacent region sequences of 122 Amerindian mtDNAs obtained from a sample set representative of the current Puerto Rico population as well as restriction data, to put forward some hypotheses on pre-Columbian female migrations to the island and population expansions occurring thereafter. Using median network analysis, I identify and describe signature sequences of nineteen maternal lineages. On the basis of genetic diversity within these sequences, nine of the lineages, constituting 84.1 percent of all Amerindian mtDNAs in Puerto Rico, show characteristics consistent with an arrival to the island in pre-Columbian times. On the same basis, I divide pre-Columbian times into three wide migration periods. I propose that four of the lineages arrived during the Archaic (Pre-Arawak) Age. I propose that three other lineages arrived substantially later, approximately by the time at which radiocarbon testing has led archaeologists to estimate the arrival of Arawaks, and that all of them underwent population expansions shortly after these migration processes. The remaining two lineages arrived later and underwent even stronger population expansions shortly after their arrival. Control region sequence and restriction analysis of the coding region identify both lineages of the third migration period and one of the second as having a South American continental origin.

## Resumen

Debido a que el modo de herencia del ADN mitocondrial (ADNmt) es estrictamente materno y no-recombinatorio, el rápido ritmo evolutivo de su región control permite una acumulación rápida de mutaciones que es útil para estudios sobre los eventos migratorios y demográficos que dieron origen a las poblaciones humanas en todo el mundo. En este capítulo, analizo secuencias de una región de 1,147 pb de longitud, que incluye la región control, en 122 ADNmts de origen amerindio que fueron obtenidos de una muestra representativa de la población puertorriqueña actual. Combino los resultados con datos de restricción para generar algunas hipótesis sobre migraciones de mujeres a la isla en tiempos precolombinos y las expansiones poblacionales que ocurrieron subsiguientemente. Usando análisis de redes medianas, identifico y describo secuencias diagnósticas de diecinueve linajes maternales. A base de la diversidad genética dentro de los linajes, nueve linajes que constituyen el 84.1 por ciento de los ADNmts amerindios en Puerto Rico muestran características consistentes con una llegada a la isla en tiempos precolombinos. A base del mismo análisis, divido los tiempos precolombinos en tres largos periodos migratorios. Propongo que cuatro de los linajes arribaron en tiempos pre-arahuacos. Propongo que tres linajes arribaron bastante más tarde, aproximadamente para el tiempo al cual las pruebas de radiocarbono han llevado a los arqueólogos a estimar la llegada de los arahuacos, y que todos ellos sufrieron expansiones poblacionales poco después de este proceso migratorio. Los restantes dos linajes llegaron más tarde y sufrieron expansiones poblacionales aún más fuertes poco después de su llegada. Análisis de restricción de la región codificante, así como de secuencias de la región control, identifican el origen continental sudamericano de ambos linajes del tercer periodo migratorio y de uno del segundo.

## Introduction

Many of the fundamental questions in Caribbean archaeology and anthropology relate to the timing and continental origin of pre-Columbian migrations to the region, as well as the extension of these migrations within the region. In theory, many of these questions could be answered using mitochondrial DNA (mtDNA) extracted from dated remains recovered under appropriately described archaeological contexts. However, ancient DNA (aDNA) analysis is a young field that presents enormous technical difficulties, especially when working with remains buried for centuries under conditions unfavorable for preservation, as is usually the case in the tropics.

A feasible alternative is to try to reconstruct the genetic history of populations using the vast amount of data that can be obtained more easily from the mtDNAs of modern populations. Because mtDNA does not recombine, modern admixed populations preserve it in a more complete state than any imaginable set of environmental conditions could preserve it in buried contexts. Hypotheses built on such reconstructions can then be tested with the analysis of mtDNA obtained from relatively few carefully selected and unearthed ancient remains.

One example in which data from modern mtDNA were instrumental in the analysis of aDNA data were those obtained in western Nevada from speakers of Numic, an Uto-Aztecan language, and from other Uto-Aztecan groups in California along with California Penutians. These data showed that the mtDNA pool of early inhabitants in western Nevada was very similar to that of modern California Penutians but significantly different from those of the modern Numic speakers inhabiting the region. Furthermore, it showed that modern Numic speakers in western Nevada held mtDNA haplogroup frequencies intermediate between the ancient inhabitants and the modern Uto-Aztecan groups. This suggested that the expanding Numic population admixed with the early inhabitants of western Nevada (Kaestle and Smith 2001).

The problem with using modern mtDNA is female mobility and the consequential uncertainty of the geographic and temporal location of such mtDNA in ancient times. This is compensated for by the large amount of modern mtDNAs that can be recovered, thus allowing the use of modern analytical methods whose power depends on sample size; these have enabled researchers to propose major migratory routes by which modern humans colonized the world after leaving Africa approximately 60,000 years ago (Forster and Matsumura 2005; Macaulay et al. 2005).

In this chapter, I perform a phylogenetic analysis of 122 mtDNAs of Amerindian origin from modern Puerto Ricans in which nineteen maternal Amerindian lineages have been identified. The phylogenetic analysis is combined with diversity analyses that suggest three major migration periods to Puerto Rico in pre-Columbian times and that these populations underwent ever more extensive expansion processes after arrival. From the incorporation of data obtained from continental mtDNAs, it is proposed that the latter two migration periods included mtDNA lineages of South American origin, whereas no indication is generated on the possible origin of the lineages belonging to the first migration period. I provide signature nucleotide sequences for each of the nineteen lineages that can be used to test the presence of them in ancient remains.

## Preferred Molecular Tools for Studies on the Genetics of Early Populations

With the advent of the polymerase chain reaction (PCR) in the 1980s (Saiki et al. 1985), the mitochondrion DNA became the molecule of choice for human evolutionary geneticists because of its strictly maternal, nonrecombinatorial mode of inheritance and its rapid evolutionary rate. Its fast mutation rate (Brown et al. 1979; Miyata et al. 1982) allowed the accumulation of high genetic diversity in the short periods of evolutionary time that elapsed between major demographic expansion or migration processes that occurred during the expansion of humans into different parts of the world (Schurr et al. 1990; Torroni et al. 1998; Watson et al. 1997). Just as importantly, the lack of recombination (Giles et al. 1980; Hutchison et al. 1974) made it possible to trace the sequence of evolutionary events occurring in this DNA. Thus, ancestral mtDNAs could be easily distinguished from derived mtDNAs, and the migration routes taken by humans could be reconstructed.

Furthermore, it was shown that some mtDNAs in founding and expanding populations often possessed specific nucleotide variants that distinguished their ancestors from all other mtDNAs in their original population. Such specific nucleotide variants became a signature for the derived population. Groups of mtDNAs sharing such variants were named "haplogroups" (Torroni et al. 1993), the specific nucleotide variants of which are often detected by Restriction Fragment Length Polymorphism (RFLP) analysis. Haplogroups tend to be geographically constrained (Cann et al. 1987; Denaro et al. 1981; Johnson et al. 1983), and thus their finding in a mixed population such as that of Puerto Rico is evidence for the migration of females from their continent of origin.

As with all phylogenetic groups, the age of haplogroups can be estimated by the genetic diversity contained within. Because the mtDNA does not re-combine, only mutations in it create new alleles, also known as haplotypes. Because the accumulation of mutations is a function of time, the conse-quential formation of new haplotypes within a haplogroup (and the re-sulting haplotypic diversity) is a measure of haplogroup age. For example, the African-specificity of haplogroups containing the highest haplotypic diversity became one of the strongest lines of molecular evidence suggest-ing an African origin for humans. In consistency with the younger age of New World populations, haplogroups specific to the New World exhibit reduced haplotype diversity. Similarly, the approximate date for the lin-eages that arrived to Puerto Rico in pre-Columbian times can be estimated based on their diversity in the island.

The human mtDNA is a closed double helix first sequenced by Ander-son and colleagues (1981) as a 16569 bp long molecule that was later cor-rected to 16568 (Andrews et al. 1999). It contains thirty-seven contiguous genes (thirteen protein-coding, two ribosomal RNA (rRNA), and twenty-two transfer RNA (tRNA) genes) in its coding region and a 1122 bp long noncoding region known as the control, or D-loop region. Because of its weaker selective constraints, the divergence rate of the control region is an order of magnitude faster than that of the coding region. Phylogenetic approaches have measured its nucleotide substitution rate at $9.883 \times 10^{-8}$ substitutions per nucleotide per year, whereas that of the coding region has been estimated at $1.068 \times 10^{-8}$ (Soares et al. 2009). Thus, a great deal of nucleotide variation information can be derived from sequencing this small region. Two hypervariable regions (HVRs) within the control region have been identified (Greenberg et al. 1983; Vigilant et al. 1989). HVR-I, located around the origin of replication, spans positions 16051 to 16400 and accumulates one nucleotide substitution every 16,677 years. HVR-II, which is found within the 3' region of the displacement loop, extends from position 68 to 263 and accumulates one substitution every 22,388 years (Soares et al. 2009). In this chapter, we combine both regions, which to-gether gain one substitution every 9,558 years. Transitions constitute 96 percent to 97 percent of base substitutions.

A great deal of variation in mutation rate exists among nucleotide sites within HVR-I and HVR-II. Some sites, classified as hypermutable, have substitution rates an order of magnitude higher than others (Wakeley 1993; Meyer et al. 1999). Thus, a weighting scheme is often implemented when analyzing HVR sequence variation, assigning less weight to transitions at hypermutable sites than to transversions or to transitions at more conserved sites.

However, only female population history can be studied through mtDNA. Hence, great efforts have been made to find genetic variation in the slowly evolving, strictly paternally inherited, nonrecombinatorial region of the Y chromosome (NRY) that could aid in deciphering male migratory history (Cruciani et al. 2002; Hammer et al. 1997, 2000, 2001; Lell et al. 2002; Santos, Bianchi et al. 1996; Seielstad et al. 2003; Underhill et al. 1996, 2000). New technological advances such as denaturing high-performance liquid chromatography has allowed the rapid identification of biallelic markers that, as in mtDNA, show the stepwise accumulation of mutations that leads to the construction of unambiguous evolutionary trees (Underhill et al. 1997, 2001; Y Chromosome Consortium 2002). NRY presented geneticists with a new challenge, as several NRY haplogroups were shown to be at high frequencies across wide geographic regions. Fortunately, as do all nuclear chromosomes, the Y contains many short tandem sequence repeats known as STRs or microsatellites that suffer length mutations at a very high rate and, as hypervariable regions in mtDNA, are useful to trace the most recent expansion and migration processes that usually occurred in more restricted geographic regions (Bamshad et al. 2001; Behar et al. 2003; Bortolini et al. 2003; Forster et al. 2000; Luis et al. 2004; Quintana-Murci et al. 2003; Rootsi et al. 2004; Zerjal et al. 2002; Zhivotovsky 2001; Zhivotovsky et al. 2004). Because the peopling of the Caribbean is a relatively recent process, dating back to possibly 7000 BP, the mtDNA HVRs and NRY microsatellites are expected to provide the most information.

## The Pre-Columbian Origin of Amerindian mtDNAs in Puerto Rico

Similar to other Latin American countries (Batista dos Santos et al. 1999; Carvajal-Carmona et al. 2000; Carvalho-Silva et al. 2001), Puerto Rico shows a strong sexual asymmetry in the continental origin of its gene pool. Preliminary studies on NRY performed in my lab at the Department of Biology of the University of Puerto Rico, Mayagüez Campus, by Katherine Martínez-Vargas and several undergraduate students, suggest that approximately 74 percent of all Y chromosomes in Puerto Rico are European in origin, 25 percent sub-Saharan African, and only 1 percent Amerindian. By contrast, a study based on 800 samples collected through

a sampling scheme designed for a sample set representative of the population of Puerto Rico showed the mtDNA distribution to be 61.3±3.4 percent
Amerindian, 27.2±3.1 percent sub-Saharan African, and 11.5±2.2 percent
West Eurasian (Martínez-Cruzado et al. 2005). Hence, the undertaking
of molecular population studies that will contribute to our knowledge of
pre-Columbian settlements in Puerto Rico is far more doable, although
only just as necessary, if based on mtDNA than on NRY. In this chapter,
I restrict myself to the sequence analysis of the control and flanking regions in Amerindian mtDNAs obtained from the sample set representative
of the Puerto Rico population and to a limited restriction analysis of the
coding region. Figure 3.1 shows the percent frequency of the four Amerindian haplogroups based on a total of 488 Amerindian mtDNAs that
were found among the 800 samples collected in Puerto Rico. Haplogroup
A is defined by the presence of a *Hae*III site at nucleotide position 663,
numbered as in the Cambridge Reference Sequence (CRS) (Anderson et
al. 1981; Andrews et al. 1999). Haplogroup B is defined by the deletion of
a 9 bp sequence that is usually found in two tandem copies between the
cytochrome oxidase second subunit and the tRNA[Lys] genes. This deletion
has arisen independently in several continents. However, only the Asian
and Native American versions occur in mtDNAs lacking both the 10394
*Dde*I and 10397 *Alu*I sites. Haplogroups C and D are defined by the gain of
an *Alu*I site at position 13262 and the loss of an *Alu*I site at position 5176,
respectively. Additional mutations help differentiate Asian versions from
those of the New World. The fifth Amerindian haplogroup, haplogroup X,
is circumscribed to North America and not found in Puerto Rico.

One important caveat to keep in mind in these analyses is the existence
of much historical documentation that demonstrates the importation of
Amerindian slaves to Puerto Rico from a wide geographic region spanning
from northern Mexico to Brazil during the period of Spanish domina-
tion (Anderson-Córdova 1990; Fernández-Méndez 1970: 91, 100, 117, 120).
Thus, Amerindian mtDNAs not native to Puerto Rico should be expected
in the sample set. Fortunately, such mtDNAs seem to form only a small
part of the sample set. The first line of evidence is the predominance of
Amerindian mtDNAs (61.3 percent) over sub-Saharan African ones (27.2
percent). As the number of African women brought to Puerto Rico over
the course of four centuries was much larger than that of Amerindian
women, the greater number of Amerindian mtDNAs in the sample set can
best be explained by the Puerto Rican origin of most individuals.

| Haplogroup | n | % ± SD |
|:---:|:---:|:---:|
| A | 256 | 52.46 ± 2.26 |
| B | 41 | 8.40 ± 1.26 |
| C | 174 | 35.66 ± 2.17 |
| D | 17 | 3.48 ± 0.83 |
| Total | 488 | --- |

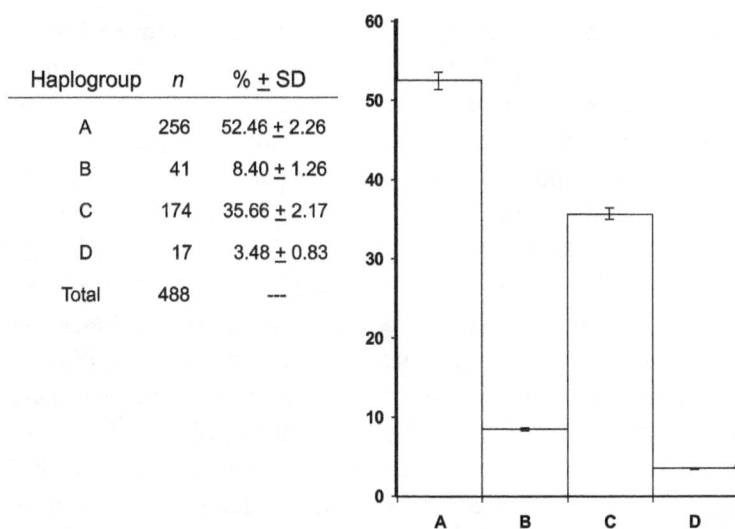

Figure 3.1. Percent frequency and standard deviation of the four Amerindian mtDNA haplogroups in Puerto Rico.

Another line of evidence is the fact that Amerindian mtDNAs in Puerto Rico as a whole exhibit relatively low gene flow characteristics. For instance, the Puerto Rico Amerindian haplogroup diversity is within the lower half of tribes in all regions of the subarctic New World, except for those regions where demographic histories were affected by strong and recent population expansion events following population replacement or bottleneck events such as the American Southwest and eastern Central America, respectively (Martínez-Cruzado et al. 2005). Furthermore, contrasting scenarios have been observed between the Andean and Amazonian regions both for the Y chromosome and for the mtDNA. Consistent with a large effective population size scenario and high gene flow between populations as supported by both the historical and the archaeological records, Andean populations show high within-population variability and an excess of rare alleles. In contrast, Amazon populations show a clinal pattern and a low frequency of rare alleles, consistent with small effective population sizes caused by isolation and restricted gene flow (Fuselli et al. 2003). The effect can be observed in table 3.1 (non–Puerto Rico data taken from Lewis et al. 2005), where the uncorrected unique HVR-I haplotype frequency of all five Andean tribes is equal to or higher than that of all Central American and Amazonian tribes. Eastern Central American tribes have gone through recent population bottlenecks followed by population expansion events and consequently have relatively low unique allele frequencies (Batista et al. 1995; Kolman et al. 1995; Kolman and Bermingham 1997; Santos et al. 1994), although usually not as low as Amazonian tribes. It is generally agreed that the Amerindian population of Puerto Rico suffered a rapid decline upon Spanish colonization, and there are few doubts about the population expansion that occurred thereafter. Thus, barring significant migration to Puerto Rico in post-Columbian times, we might expect the Puerto Rico Amerindian uncorrected unique HVR-I haplotype frequency to be similar to those of eastern Central American tribes, which it is. As unique haplotype frequencies can be sensitive to sample size, I adjusted the sample size to 36 (the average of all tribes in table 3.1), and found the Puerto Rico frequency to be lower than two of the three eastern Central American tribes. When the sample size is adjusted to the smallest and largest sample sizes in the table, the calculated frequencies are 0.583 for $n = 22$ and 0.412 for $n = 63$, always lower than at least one of the three eastern Central American tribes and all five Andean tribes.

As the uncorrected unique HVR-I haplotype frequency tends to decrease with sample size except when the sample size is very small, another approach to this analysis is to correct the unique haplotype frequency by multiplying it by the harmonic mean of the sample size. Table 3.1 shows this correction for all tribes, including Puerto Rico with $n = 122$. The use of this approach only raises the Ngöbé over two Andean tribes and raises Puerto Rico (n = 122) over one Amazonian and one Central American tribe. In summary, the Puerto Rico unique haplotype frequency can be described as similar to those of eastern Central American tribes regardless of the sample size used. It is consistent with that of a population undergoing an expansion after a bottleneck event and not with that of a population having high gene flow with other populations (as exhibited by Andean tribes) or as would be expected if it was composed mostly of recent migrants of various origins.

An additional line of evidence for the pre-Columbian origin of most modern Amerindian mtDNAs in Puerto Rico is that they form concise phylogenetic groups upon median network analyses of their HVR sequences, and not a plethora of isolated haplotypes as may have been expected if they represented mostly post-Columbian migrations from diverse continental regions. Nine polymorphic phylogenetic groups or lineages compose 84.1 percent of all Amerindian mtDNAs in Puerto Rico (see below). The remaining 15.9 percent is distributed among ten monomorphic phylogenetic

Table 3.1. Unique HVR-I Haplotype Frequency in Puerto Rico and in Central and South American Tribes

| Tribe | $n$ | Number of HVR-I haplotypes | Uncorrected unique HVR-I haplotype frequency | Corrected unique HVR-I haplotype number | Region |
|---|---|---|---|---|---|
| Mapuche | 39 | 14 | 0.786 | 3.322 | Andes |
| Tayacaja | 61 | 42 | 0.714 | 3.343 | Andes |
| Ancash | 33 | 27 | 0.704 | 2.856 | Andes |
| Arequipa | 22 | 18 | 0.667 | 2.430 | Andes |
| Ngöbé | 46 | 8 | 0.625 | 2.747 | Central America |
| Cayapa | 30 | 8 | 0.625 | 2.476 | Andes |
| Huétar | 27 | 7 | 0.571 | 2.203 | Central America |
| Puerto Rico[a] | 36 | 15 | 0.467 | 1.935 | Caribbean |
| Gavião | 27 | 7 | 0.429 | 1.652 | Amazon |
| Zoró | 28 | 8 | 0.375 | 1.459 | Amazon |
| Puerto Rico | 122 | 21 | 0.286 | 1.536 | Caribbean |
| Kuna | 63 | 7 | 0.286 | 1.346 | Central America |
| Xavante | 25 | 4 | 0.250 | 0.944 | Amazon |

[a] Sample size adjusted from 122 to 36.

groups, most of which probably represent the expected Amerindian mtD-NAs not native to Puerto Rico.

## Methods

In this chapter, I present, in the form of median networks, a total of 122 modern Amerindian mtDNA sequences from Puerto Rico spanning positions 15924 to 501, including the last 30 bp of the tRNA$^{Thr}$ gene, the complete 69 bp long tRNA$^{Pro}$ gene, 1 bp separating these two genes, and 1047 bp of the control region, including both HVR-I and HVR-II. The sequence information is supplemented with some coding region RFLP analysis. I describe signature sequences of nineteen mtDNA lineages and evidence to support the pre-Columbian arrival to Puerto Rico of nine of these lineages. The nine lineages encompass 84.1 percent of all Amerindian mtDNAs currently in Puerto Rico. The arrival and date of arrival of these lineages can now be tested by searching their signature sequences in dated ancient remains.

A single unmodified median network contains all most-parsimonious phylogenetic trees within a haplogroup and graphically displays all sequence data used for its construction (Bandelt et al. 1995, 2000). The diagram in figure 3.2 is an example of how a median network displays sequence data. Sequence haplotypes are represented by circles, the relative size of which reflects their frequency. In this example, there is a sample size of seven with three sequence haplotypes. Haplotype I is the most frequent ($n = 4$), and the area of its circle is four times larger than that of haplotype

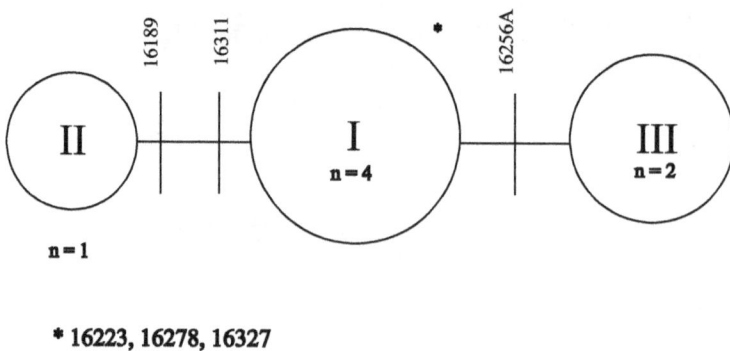

* 16223, 16278, 16327

Figure 3.2. Diagram of a median network.

II, a unique haplotype. The sequence of the central haplotype is indicated by stating at the bottom of the network all nucleotide sites showing differences with the CRS. Haplotypes I and II differ by two transitions that are indicated by crosshatches and identified by specifying the nucleotide site where they occur. When the nucleotide difference is a transversion, the identity of the transversion is indicated by specifying the base of the experimental sequence. Thus, in figure 3.2, haplotype I differs from the CRS by three transitions, haplotype II by five, and haplotype III by three transitions and one transversion. In addition, because they give rise to all other haplotypes, founding haplotypes are expected to be located at the center of the network. Furthermore, because they are the oldest of all haplotypes in their respective lineages, barring population bottlenecks, they are expected to have the highest frequency.

## Analysis and Discussion

The median network constructed with thirty-nine haplogroup C samples is shown in figure 3.3. The PCR amplification of the fragments and their purification were performed in the lab by Eimy González-Bonilla and Elizabeth Guzmán-Morales. DNA sequencing was serviced by the University of Medicine and Dentistry at the New Jersey Molecular Resource Facility (www.umdnj.edu/mrfweb). Length mutations in the C-runs spanning positions 16184 to 16193 and 303 to 315 were ignored. For the number of mutations between them, four lineages can be identified. Lineage C-I is the most common Amerindian mtDNA lineage in Puerto Rico. As explained just below, this conclusion is based on RFLP analysis.

Torroni and colleagues (1993) performed high-resolution RFLP analysis of the entire mtDNA in 301 Amerindians belonging to two tribes from North America, five from Central America, and nine from South America. Of these Amerindians, 61 belonged to haplogroup C and were distributed through twenty-five RFLP haplotypes with thirty polymorphic restriction sites. Twenty-one of these haplotypes were private (that is, found in only one of the sixteen tribes examined). Two others, AM79 and AM82, were found in two Amazonian tribes each. The remaining two haplotypes were AM32 and AM43. The first was found in one North American and four South American tribes. The second was found in one tribe each of Central and South America. These two haplotypes differed by only one restriction site: AM32 lacked the 16517 HaeIII site whereas AM43 had it. This site is

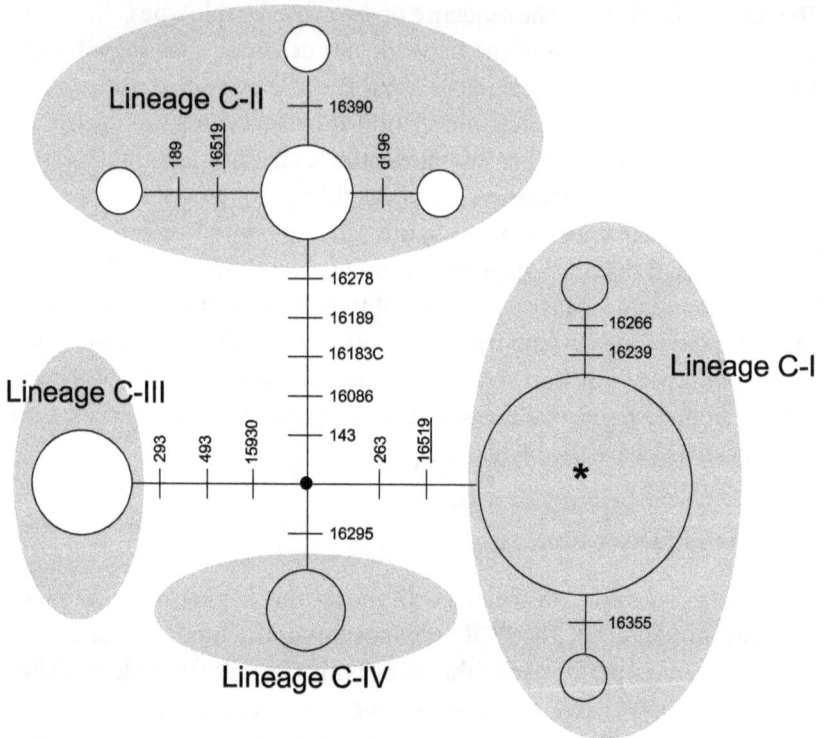

* 16223, 16298, 16325, 16327, 16519, 73, d248-249, 2d286-291, 489, 493

Figure 3.3. Median network constructed with thirty-nine haplogroup C mtDNAs. Haplotypes are represented by open circles. The smallest open circles represent unique haplotypes. The black, small circle represents a hypothetical ancestor in the continent. Mutations listed at the bottom of the figure refer to deviations from the CRS of the marked haplotype (*). Mutations between haplotypes are specified in crosshatches by the number of their position according to the CRS. The order of mutations in uninterrupted branches is arbitrary. The designation *d196* implies a 1 bp deletion at position 196. Underlines denote recurrent mutations. The 7013 *Rsa*I mutation falls in the branch containing transitions 263 and 16519. Transition 16519 alters the 16517 *Hae*III site.

highly variable; it has been found polymorphic in most, if not all, world haplogroups and thus is of little informational value. Based on haplotype frequency and geographic distribution, Torroni and colleagues (1993) chose these two haplotypes as the probable founders of the New World. However, the central location of AM32 in the radiation of its haplogroup as observed in phylogenetic analyses suggested it may be the haplogroup C founder of the New World.

More recently, complete mtDNA sequence analyses have generated a

worldwide phylogeny of mtDNAs, including the emergence in Asia of haplogroup C from the CZ clade. Haplogroup C has now been divided into
four subhaplogroups, named C1 to C4 (Volodko et al. 2008). AM32 corresponds to C1, which contains all haplogroup C mtDNAs in the New World
except for a small subhaplogroup named C2c that was found recently in an
Ijka-speaking Colombian tribe (Tamm et al. 2007). AM43 only represents
a heterogenous group of subclades derived from AM32 through repeated
16519 transitions occurring in parallel and creating the 16517 HaeIII site
that distinguishes it from AM32.

Analyses of complete Native American mtDNA sequences through
Bayesian Skyline Plots suggest that the founder population of the New
World spent a few thousand years in isolation in Beringia and suffered
a small reduction in size before expanding strongly into the New World
at 15,000 to 20,000 BP (Fagundes et al. 2008). By the time of this expansion, C1 was already subdivided into the Native American–specific subhaplogroups C1b, C1c, and C1d, leaving subhaplogroup C1a in Asia. C1b
mtDNAs are characterized by a transition at position 493. Those belonging
to C1c have transitions at positions 1888 and 15930, while C1d mtDNAs
exhibit transitions at 7697 and 16051 (Achilli et al. 2008).

Our first paper on this subject (Martínez-Cruzado et al. 2001) presented
a survey made in western Puerto Rico from which twenty-one haplogroup
C samples were identified. Twenty of these possessed the 16517 HaeIII site.
Mayra Troche-Matos (2001) took three of these plus the one lacking the
site and tested them for five additional restriction sites informative for
South American origin on the basis of the study by Torroni and colleagues
(1993). She found that the three mtDNAs having the 16517 HaeIII site
lacked the 7013 RsaI and the 3391 AluI sites but possessed the 3412 HaeIII
site, thus identifying them as belonging to haplotype AM79, which Torroni
and colleagues (1993) had found only in two Amazonian tribes: the Kraho
and the Yanomami. Later, Alicia Román-Colón, Juan Ramírez-Lugo, and
I tested all 174 haplogroup C samples from the set representative of Puerto
Rico for the 7013 RsaI and 16517 HaeIII sites and found 104 of them having
a -/+ motif, 66 +/-, and 4 +/+. This was followed by an analysis by Jennifer Startek, Román-Colón, and Ramírez-Lugo, who tested 31 -/+, 30 +/-,
and 1 +/+ samples for 15 additional polymorphic restriction sites designed
to unambiguously identify their RFLP haplotype.[1] All samples presented
the exact same pattern for the 15 restriction sites: the 31 -/+ samples were
confirmed as belonging to haplotype AM79, the 30 +/- were AM32, and
the +/+ was classified as AM43. It was thus reasonable to extrapolate and

conclude that the 104 samples showing the -7013 *Rsa*I / +16517 *Hae*III motif belonged to RFLP haplotype AM79, the 66 with the +7013 *Rsa*I / -16517 *Hae*III motif were AM32, and the remaining 4 were AM43.

Thirty-nine haplogroup C samples were randomly chosen to construct the median network under the condition of representative proportion of the RFLP haplotypes. Thus, 24 (61.5 percent) of these haplogroup C samples were AM79, 14 (35.9 percent) were AM32, and 1 (2.6 percent) was AM43. The median network revealed four lineages, three of which possess the transition at position 493 relative to the CRS that identifies them as belonging to the complete mtDNA sequence subhaplogroup C1b (figure 3.3). Lineage C-I contains all and only the 24 AM79 samples. That is, there is a one-to-one correspondence between those samples belonging to RFLP haplotype AM79 and those belonging to lineage C-I. That lineage C-I corresponds to RFLP haplotype AM79 is supported by the work of Achilli and colleagues (2008) through complete mtDNA sequencing. They found that AM79 exhibits the transition at 493 that distinguishes C1b mtDNAs, and they assigned it the name C1b2. Figure 3.3 shows that lineage C-I is a C1b lineage, as it possesses the transition at position 493. As 104 out of the 488 Amerindian mtDNAs belong to AM79, AM79 and lineage C-I must represent 21.31 percent of all Amerindian mtDNAs in Puerto Rico. No other lineage approaches that percentage (see below).

As both the -7013 *Rsa*I motif and haplotype AM79 have been found only in the Amazon, it is reasonable to conclude that lineage C-I represents a South American migration to Puerto Rico. In addition, lineage C-I meets the criteria for a starlike phylogeny. Starlike phylogenies have a central, founder haplotype more frequent than those derived from it. They are defined by $S > \pi > 1.5\rho$ where $S$ represents the number of segregating sites, $\pi$ the average pairwise number of differences between samples, and $\rho$ the average number of differences between any sample and the founding haplotype where the founding haplotype can be identified because it occupies a central position in the starlike phylogeny (Forster et al. 1996). Meeting starlike phylogeny criteria is an indication of a population expansion soon after the arrival of the lineage. It is thus apparent that lineage C-I underwent rapid population expansion soon after its arrival to Puerto Rico. The time of such population expansion can also be estimated by using $\rho$ as a molecular clock where one unit, measured using only HVR-I and HVR-II, is equivalent to 9,558 years (Soares et al. 2009; standard deviation calculated as in Saillard et al. 2000). I estimate the population expansion to have occurred 1195±690 BP. Thus, lineage C-I represents a migration process from South America during the Arawak Age.

Although much less frequent than C-I, lineage C-II presents a similar history. It was recently found by our graduate student, Marcela Díaz-Matallana, in a Venezuelan from Caracas living in Puerto Rico. It has also been found in a subject from northwestern Venezuela (Dinorah Castro, personal communication). It meets the criteria for a population expansion upon arrival to Puerto Rico, and its time of arrival is estimated at 2731±1931 BP. Thus, it may represent a pre-Columbian migration to Puerto Rico from South America preceding or concurrent with the C-I migration. The complete mtDNA sequence name of lineage C-II is C1b4. No complete mtDNA sequence names for lineages C-III and C-IV can be identified from the work of Achilli and colleagues (2008). However, the transition at position 493 identifies C-IV as a C1b lineage, while the lack of this transition combined with a transition at position 15930 shows that C-III is a C1c lineage.

It is important to note that although lineages C-I and C-III differ by four sites in the control region, one site inside the tRNA$^{Thr}$ gene, and the 7013 RsaI site, their HVR-I haplotype (16223, 16298, 16325, 16327) is identical among themselves and to that identified by Lalueza-Fox and colleagues (2001) as the one and only haplogroup C founder in the Dominican Republic. The latter work, based on ancient remains of dates ranging from AD 670±70 to AD 1680±100 and collected in the La Caleta archaeological site, was limited to HVR-I sequences. Evidently, the exploration of HVR-II sequences in ancient remains belonging to haplogroup C will be critical to distinguish between lineages. The data suggest that it will also be sufficient, as three of the four lineages are characterized by particular HVR-II mutations relative to their hypothetical common ancestor (figure 3.3).

Another interesting aspect of haplogroup C lineages in Puerto Rico is their geographic distribution. From samples taken in western Puerto Rico, Martínez-Cruzado and colleagues (2001) found that twenty out of twenty-one haplogroup C samples carried the +16517 HaeIII motif. However, the study based on a sample set representative of the population of the island found that mtDNAs carrying the +16517 HaeIII motif accounted for only 62.1 percent of all haplogroup C samples. This suggests that haplogroup C mtDNAs carrying the -16517 HaeIII motif, which account for almost all mtDNAs belonging to lineages C-II, C-III, and C-IV, may be rare in western Puerto Rico. The sample collection procedure in the latter study included a questionnaire where participants were asked for the municipality

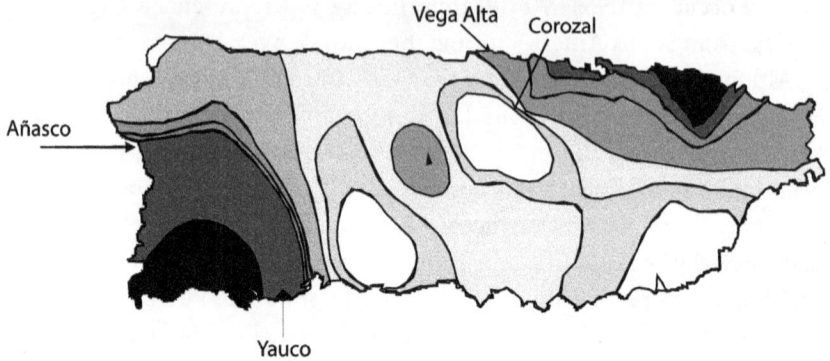

Figure 3.4. Frequency-based contour map constructed with Map Viewer 7 (Golden Software, Inc.) plotting -7013 *RsaI* (lineage C-I) versus +7013 *RsaI* (lineages C-II, -III, and -IV) haplogroup C mtDNAs in recent maternal ancestors of study participants in Puerto Rico. Darkest gray indicates highest -7013 *RsaI* frequency; white, highest +7013 *RsaI* frequency. The approximate location of the municipalities of Yauco, Añasco, Vega Alta, and Corozal are indicated. Of the 174 haplogroup C mtDNAs, only one, belonging to the +7013 *RsaI* type, was traced back to outside of Puerto Rico; it is not included in the analysis.

in which the family of mothers, maternal grandmothers, and maternal great-grandmothers lived at their time of birth. This information was used to display the geographic distribution of haplogroup C mtDNAs within Puerto Rico approximately two generations ago in figure 3.4. In 95 percent of the cases, the sample was traced back to a date earlier than 1945, before the occurrence of massive migrations within the island after World War II. Figure 3.4 indeed shows a depletion of mtDNAs belonging to the aggregate of lineages C-II, C-III, and C-IV in the southwest. There are twenty-two haplogroup C samples in the region from Yauco to Añasco, and twenty-one of them belong to lineage C-I. Because no excess of haplogroup C has been found in this region (Martínez-Cruzado et al. 2005), this result suggests a lack of +7013 *RsaI* (C-II, C-III, and C-IV) lineages rather than an excess of lineage C-I in southwestern Puerto Rico. The opposite could be occurring in another region of the island. Twelve haplogroup C samples were traced back to the municipalities of Corozal and Vega Alta, which cover most of the Cibuco River basin, and only one of them belonged to lineage C-I. It is uncertain whether this is the result of the lack of lineage C-I or an excess of the aggregate of the other three lineages because of the small sample size. However, Corozal was one of only three municipalities

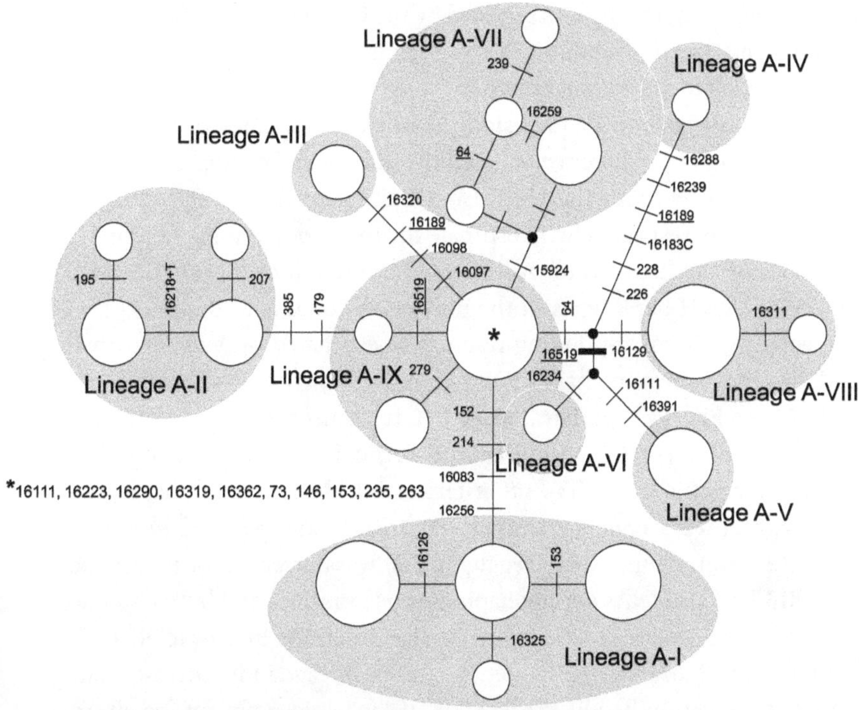

Figure 3.5. Median network constructed with fifty-one haplogroup A mtDNAs. Mutations, haplotypes, and hypothetical intermediate haplotypes are indicated as in figure 3.3. The order of mutations in uninterrupted branches is arbitrary, and recurrent mutations are highlighted by underlines. Transition 16519 alters the 16517 HaeIII site. Parallel lines in a reticulation (lineage A-VII) denote that the mutated position is identical. A16218+T in a branch within lineage A-II indicates that the 16218 transition is always accompanied by a triplication of the 9 bp sequence, commonly found duplicated between the cytochrome oxidase second subunit and the tRNA$^{Lys}$ genes. Mutations listed at the bottom of the figure refer to deviations from the CRS of the marked haplotype (*).

out of the twenty-eight sampled in which haplogroup C was recovered with a frequency higher than haplogroup A, and an excess of +7013 RsaI lineages is thus likely.

The thirty-nine haplogroup C samples were distributed among nine haplotypes in four lineages. Haplogroup A is more diverse, with the fifty-one samples in the median network (figure 3.5) distributed among twenty-one haplotypes and apparently nine lineages. To construct the haplogroup A median network, the PCR amplification of the fragments and their purification were performed by Michelle Rivera-Vega and Patricia Valencia-

Rivera. DNA sequencing was serviced by the UMDNJ Molecular Resource Facility, and length mutations in the C-runs spanning positions 16184 to 16193 and 303 to 315 were ignored.

To estimate with more precision the relative frequency of haplogroup A lineages, Valencia-Rivera estimated the fractions of haplogroup A mtD-NAs with and without the 16517 *Hae*III site gain: 21 and 99 out of 120 samples tested, respectively. To estimate the frequency among Amerindian mtDNAs of any haplogroup A lineage, the fraction of the +16517 *Hae*III or -16517 *Hae*III haplotypes in the network belonging to the lineage was corrected by its corresponding fraction as estimated by Valencia-Rivera and by the fraction of haplogroup A mtDNAs among all Amerindian mtDNAs in Puerto Rico. For lineage C-I, the estimate was more direct, as exactly 104 of the 488 Amerindian mtDNAs collected were haplogroup C mtDNAs meeting the -7013 *Rsa*I criterion. For the remaining haplogroup C lineages, their frequency among +7013 *Rsa*I haplogroup C lineages as estimated from the network (see figure 3.3) was corrected for the fraction of +7013 *Rsa*I mtDNAs among haplogroup C samples and by the fraction of this haplogroup among all Amerindian mtDNAs in Puerto Rico. For haplogroups B and D, the frequency of each lineage in its corresponding network was multiplied by the fraction of the haplogroup among all Amerindian mtDNAs. The results are shown in table 3.2.

The very low average number of mutations commonly separating nearest haplotypes in the haplogroup A median network suggests that two mutations between nearest haplotypes may be indicative of separate origins. By assigning those haplotypes differing among themselves by more than one mutation to different lineages, I identified nine lineages (figure 3.5). Lineage A-I can easily be distinguished from all other haplogroup A mtD-NAs by signature mutations in both HVR-I (16083, 16256) and HVR-II (152, 214). This set of mutations has not been found outside of Puerto Rico, and the continental origin of lineage A-I cannot be determined. Its time of arrival can also not be estimated, as the lineage does not fit the starlike phylogeny criteria. Lineage A-I is the most frequent among haplogroup A lineages (table 3.2). The diversity within the lineage suggests a relatively early arrival to Puerto Rico.

As with some haplogroup C lineages, lineage A-II can be identified only through control region sequences outside of HVR-I. Its signature motif is 179 and 385, and (together with lineages C-II, A-I, and A-VII) this lineage holds the distinction of having the largest number of haplotypes: four. This

Table 3.2. Lineage Parameters

| Lineage | Frequency[a] | Haplotype number | $h$[b] | $S$[c] | $\Pi$[d] | Signature sequence[e] |
|---------|-----------|------------------|--------|--------|--------|------------------------|
| A-I | 16.38 | 4 | 0.7582 | 0.9434 | 1.0769 | 16083, 16256, 152, 214 |
| A-II | 5.65 | 4 | 0.7857 | 1.1570 | 1.0714 | 16519, 179, 385 |
| A-III | 2.34 | 1 | -- | -- | -- | 16097, 16098, 16189, 16320 |
| A-IV | 1.17 | 1 | -- | -- | -- | 16183C, 16189, 16239, 16288, 64, 226, 228 |
| A-V | 2.12 | 1 | -- | -- | -- | 16111, 16391, 16519, 64 |
| A-VI | 0.71 | 1 | -- | -- | -- | 16234, 16519, 64 |
| A-VII | 7.02 | 4 | 0.8000 | 1.3139 | 1.2667 | 15924 |
| A-VIII | 7.02 | 2 | 0.2857 | 0.4082 | 0.2857 | 16129, 64 |
| A-IX | 10.06 | 3 | 0.5556 | 0.7359 | 0.6111 | -- |
| B-I | 6.92 | 2 | 0.1429 | 0.3145 | 0.1429 | -- |
| B-II | 0.49 | 1 | -- | -- | -- | 189, 271 |
| B-III | 0.49 | 1 | -- | -- | -- | 271, 295 |
| B-IV | 0.49 | 1 | -- | -- | -- | 150, 152, 185, 189, |
| C-I | 21.31 | 3 | 0.1630 | 0.8034 | 0.2500 | -- |
| C-II | 6.69 | 4 | 0.7143 | 1.6327 | 1.1429 | 16086, 16183C, 16189, 16278, 16519, 143, 263 |
| C-III | 4.78 | 1 | -- | -- | -- | 15930, 16519, 263, 293, 493 |
| C-IV | 2.87 | 1 | -- | -- | -- | 16295, 16519, 263 |
| D-I | 3.05 | 3 | 0.7143 | 0.6289 | 0.9341 | -- |
| D-II | 0.44 | 1 | -- | -- | -- | 16142, 16179, 16295, 16497 |

[a] Relative to Amerindian mtDNAs in Puerto Rico.
[b] Heterozygosity, the probability that two samples taken at random belong to different haplotypes (Tajima 1989).
[c] Number of segregating sites (Forster et al. 1996).
[d] Pairwise number of differences (Li 1997).
[e] Relative to the sequence of the marked (*) haplotype in the respective network.

high diversity comes in spite of its relatively low frequency; only eight mtDNAs compose this lineage (compared to fourteen for lineage A-I), and this may be indicative of a very early arrival. Lineage A-II could be divided into two sublineages, as the four mtDNAs with the 16218 transition also exhibit a triplication of the 9 bp sequence located between the cytochrome oxidase second subunit and the tRNA$^{Lys}$ genes that is commonly found duplicated (the presence of which in only one copy constitutes the specific variant for haplogroup B). This triplication has been observed previously. Four of the six Amerindian mtDNAs obtained from Indiera Alta in Maricao by Martínez-Cruzado and colleagues (2001) belonged to haplogroup A and had the +16517 HaeIII motif, as do all members of lineage A-II. Three of these four mtDNAs also exhibited the 9 bp triplication. I sequenced the

HVR-I of one of these three mtDNAs and confirmed the presence of the 16218 transition. Thus, lineage A-II, especially the sublineage with the triplication, seems to be very common in Indiera Alta, in spite of the fact that lineage A-II may represent only 5.7 percent of all Amerindian mtDNAs in Puerto Rico (see table 3.2). As the Amerindian heritage of Indiera Alta probably stems from Mona Island (Dávila-Dávila 2003: 36), lineage A-II must be regarded as a prime candidate to represent the ancient population of that island.

Lineage A-III most probably represents a post-Columbian migrant. Haplotype BR2 from southeastern Brazil has its exact same control region motif (16097, 16098, 16189, 16320), except that it has an additional transition at position 16142 (Alves-Silva et al. 2000). The lines of evidence for the post-Columbian origin of lineage A-III are its low frequency and its total lack of diversity. Lineages A-IV, A-V, and A-VI are also monohaplotypic and of a frequency similar to A-III and are thus likewise regarded as probable post-Columbian migrants.

I have recently been able to separate lineages A-VII, A-VIII, and A-IX because of the extension of most of the sequences determined into the tRNA$^{Thr}$ gene. These have shown the existence of a transition at position 15924 in only six samples that led me to conclude that position 64, which is hypermutable in haplogroup A mtDNAs, has mutated twice in the haplogroup A network of Puerto Rico. Thus, lineage A-VII arises as a lineage with a transition at position 15924 as its signature. It is slightly more diverse than lineage A-II and thus may represent a very early arrival to Puerto Rico. It fits the starlike phylogeny criteria, and its time of arrival is estimated at 7965±3562 BP, clearly under the Pre-Arawak Age. Lineage A-VII corresponds to the complete mtDNA sequence subhaplogroup A2k1 (Achilli et al. 2008). The only other mtDNA from outside Puerto Rico belonging to this subhaplogroup was observed in a Wayuu Indian from the La Guajira Peninsula, and it might suggest that the very first people to arrive to Puerto Rico were South Americans as well.

In spite of the fact that the main haplotypes of lineages A-VIII and A-IX only differ between themselves by two transitions at highly variable sites (Soares et al. 2009), I regard them as two distinct lineages. These two haplotypes are the only haplogroup A haplotypes shared with the Dominican Republic (Feliciano-Vélez 2006). In that country, as in Puerto Rico, the hypothetical intermediate haplotype between these two haplotypes is missing. As the disappearance of this same and only haplotype in both countries would be very hard to explain (especially given the fact that, if it existed, its nearest haplotypes would be the most common in both countries), it is very likely that this haplotype never existed in these islands. I thus conclude that A-VIII and A-IX represent distinct female lineages migrating to the Greater Antilles

Both lineages meet the criteria for a population expansion soon after arrival. The time of arrival to Puerto Rico is estimated at 2048±1365 and 2130±2130 for lineages A-VIII and A-IX, respectively. Because of the large margins of error, it cannot be stated categorically if they arrived in Arawak or Pre-Arawak times. Their continental origin cannot be ascertained either, as the founder haplotypes of these two lineages are very common in North America, as well as in Central and South America. It cannot even be ruled out that these lineages may have arrived to both islands independently. However, the fact that these two lineages possess the two most common founder haplogroup A haplotypes in both countries suggests either a pre-Columbian arrival that spread across the Greater Antilles, or the arrival of these haplotypes in substantial numbers from continental locations where they were in high frequency.

Enid Gómez-Sánchez and Vimalier Reyes-Ortiz amplified and purified the fragments corresponding to seventeen haplogroup B mtDNAs and sixteen haplogroup D mtDNAs, the sequences of which were obtained as above and used to construct the corresponding networks. One haplotype contained thirteen of the seventeen haplogroup B mtDNAs, and the remaining four mtDNAs were distributed through four unique haplotypes (figure 3.6). Remarkably, all differences among haplotypes reside outside the HVR-I. The HVR-I motif that all haplotypes share is 16092, 16182C, 16183C, 16189, 16217, 16249, 16312, 16344. Following the criterion that more than one mutation between haplotypes may suggest independent origins, I have divided the haplogroup into four lineages. Lineages B-II, B-III, and B-IV are monomorphic and thus probably represent post-Columbian migrations to Puerto Rico. This idea is supported by their distribution inside Puerto Rico, as the maternal ancestors of the participants carrying these mtDNAs were all from the northwest corner of the island. By contrast, lineage B-I is a relatively frequent lineage (table 3.2) composed of two haplotypes, and I thus consider it to represent a pre-Columbian, relatively recent migration to Puerto Rico.

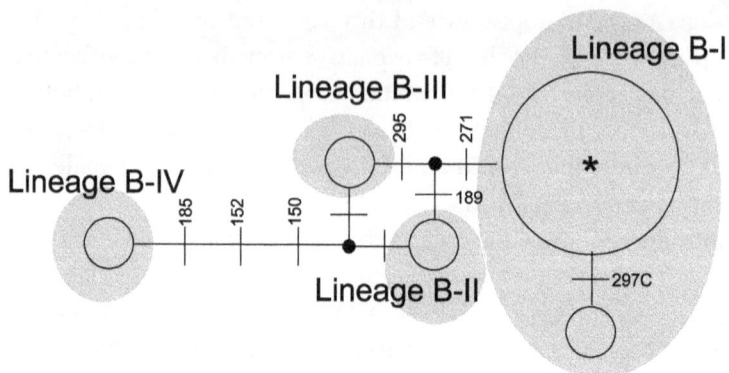

*16092, 16182C, 16183C, 16189, 16217, 16249, 16312, 16344, 16519, 73, 152, 263, 271

Figure 3.6. Median network constructed with seventeen haplogroup B mtDNAs. Mutations, haplotypes, and hypothetical intermediate haplotypes are indicated as in figure 3.3. The order of mutations in uninterrupted branches is arbitrary. Mutations listed at the bottom of the figure refer to deviations from the CRS of the marked haplotype (*).

It is interesting that despite their apparent disparate origins and the considerable diversity among them, all haplogroup B mtDNAs share a common HVR-I motif, especially because this motif is not common in the literature. Only two samples in the literature share the last five transitions of this motif, and both are found in northern and northeastern Brazil (Alves-Silva et al. 2000; haplotypes BR27 and BR28). This suggests that the HVR-I motif is predominant in the continental subregion from which all these samples originated and that it has been in that subregion long enough to accumulate the diversity observed. One haplotype made it to Puerto Rico in pre-Columbian times, giving rise to lineage B-I. Other members from that big continental family, and not from any other haplogroup B family, made it to Puerto Rico in post-Columbian times and gave rise to the remaining lineages. The prime candidate for this continental subregion is northern South America.

The only mutation in lineage B-I, at position 297, lies outside HVR-I and HVR-II. Thus, to estimate its time of arrival we used the mutation rate calculated for the entire control region, which stands at 1 mutation every 9,058 years (Soares et al. 2009). This may lead us to a slight underestimation of the time of arrival for lineage B-I, as 75 bp of the 1122 bp control region were not included in our analysis. Lineage B-I was estimated to have arrived to Puerto Rico 647±647 BP, very late in pre-Columbian times.

The sixteen haplogroup D samples are distributed among four haplotypes, three of which contain fourteen of the samples and constitute lineage D-I (figure 3.7). The central haplotype of this lineage seems to be widespread both in Central America (Santos et al. 1994; Torroni et al. 1993) and in South America (Alves-Silva et al. 2000; Ward et al. 1996). Its HVR-I motif has also been found in the ancient remains of two out of five Cuban Ciboneys belonging to haplogroup D (Lalueza-Fox et al. 2003) and of two out of six Dominican Taínos of the same haplogroup (Lalueza-Fox et al. 2001). Because it is a concise, polymorphic lineage and because this lineage may have been present elsewhere in the Greater Antilles in pre-Columbian times, I regard it as native to Puerto Rico. However, lineage D-I is by far the least frequent of all lineages we consider native to Puerto Rico (table 3.2) and is less frequent than lineage C-III, a monomorphic lineage that may have arrived in post-Columbian times. Evidently, lineage D-I was not a common lineage among Puerto Rican Taínos or went through a particularly hard bottleneck upon Spanish colonization. Interestingly, the maternal ancestors of all fourteen participants bearing lineage D-I mtDNA were restricted to the eastern half

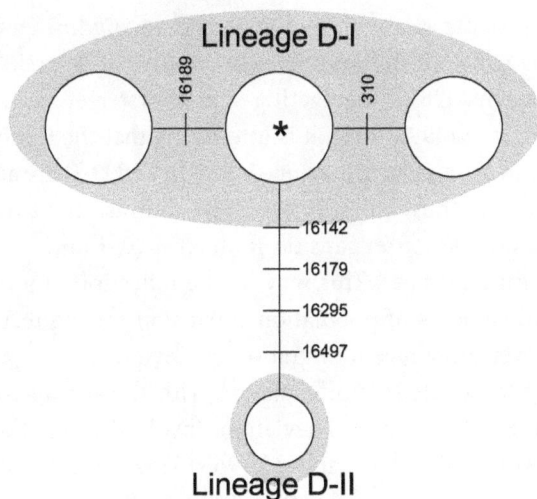

Figure 3.7. Median network constructed with sixteen haplogroup D mtDNAs. Mutations and haplotypes are indicated as in figure 3.3. The order of mutations in uninterrupted branches is arbitrary. Mutations listed at the bottom of the figure refer to deviations from the CRS of the marked haplotype (*).

of the island. This observation might be related to a political partition in the island that archaeological evidence suggests developed concomitant to a chiefdom society after 600 AD.

## Conclusions

Research conducted by myself and colleagues has identified nineteen mtDNA lineages, nine of which are proposed to have arrived to Puerto Rico in pre-Columbian times, based primarily on their diversity. These nine lineages account for 84.1 percent of all Amerindian mtDNAs in Puerto Rico. They are all (and the only ones) polymorphic. In terms of frequency in relation to all other lineages, they occupy positions one–eight and ten (table 3.2). The nine lineages regarded to have arrived to Puerto Rico in pre-Columbian times are A-I, A-II, A-VII, A-VIII, A-IX, B-I, C-I, C-II, and D-I.

An attempt to date the arrival of these lineages is stymied by the low sample size of most lineages that results in very large standard deviations, and by the lack in three of the lineages of the population expansion footprint required for dating. However, a group of diversity measures could be employed to propose relative times of arrival among lineages without specifying dates. Table 3.2 shows different measures of diversity, such as heterozyosity (Tajima 1989), segregating sites (Forster et al. 1996), and pairwise differences (Li 1997: 238). It is interesting that the three lineages with no population expansion footprint (A-I, A-II, and D-I) are among the five most diverse and, thus, oldest lineages. By contrast, the two lineages with the biggest population expansion footprints (C-I and B-I) are the youngest of the nine lineages. This may be the reflection of a pattern of continuous improvements of population expansion fueling technologies associated with ever more recent migration processes.

In an attempt to collate this information with the archaeological record, and recognizing the standard deviations involved, I hypothesize that lineage A-VII, estimated to have arrived 7965±3562 BP, and the lineages without a population expansion signature (A-I, A-II, and D-I) arrived to Puerto Rico during the Pre-Arawak period. We have no indication of the continental origin of any of these lineages. Around the Arawak migration that occurred during the last millennium BC, lineages C-II, A-VIII, and A-IX arrived, with C-II probably arriving earlier than the other two lineages. Lineage C-II has a northern South American origin. Both lineages

C-I and B-I arrived from South America later in Arawak times and underwent extensive population expansions.

In summary, this work has proposed the existence of four maternal lineages in modern Puerto Rico that arrived in Pre-Arawak times, three that arrived approximately concordant with a first Arawak migration process, plus two other lineages that arrived subsequently from South America and experienced strong population expansions upon arrival. Furthermore, it has described signature nucleotide sequences that will guide ancient DNA analyses aiming to relate the lineages identified here to an archaeological context.

This work should benefit from a larger sample size that could help reduce the enormous standard error in time estimates and raise certainty in the number of migration processes that reached the island in pre-Columbian times as well as their time of arrival. In addition, sampling in neighboring islands and continental regions must be undertaken to explore the geographic origin of the lineages that probably arrived in pre-Columbian times and their possible routes of entry, as well as to examine levels of female-mediated gene flow across the Caribbean.

## Acknowledgments

I thank all volunteers who gave hair root samples for this project. I also thank T. Arroyo-Cordero, A. Ayala-Rodríguez, H. Cerdá-Ramos, L. Colón-Negrón, J. Henry-Sánchez, R. Hernández-Rodríguez, J. Irizarry-Ramos, D. Knudson-González, M. Lacourt-Ventura, H. López-López, R. Lugo-Sánchez, F. Maldonado-Chamorro, N. Martínez-Aquino, A. Merced-Alejandro, C. Montalvo-Vélez, P. Ochoa-Méndez, J. C. Ortiz-Santiago, T. Pagán-Ortega, J. Pérez-Sepúlveda, E. Ríos-Santiago, R. Rivera-Torres, R. Rodríguez-Ramos, F. Rosas-Rivera, A. Seda-Martínez, C. Varela-Valentín, and L. Vélez-Ortiz, students of the University of Puerto Rico at Mayagüez (UPRM) who undertook the strenuous task of identifying the selected adults, educating them on the project, interviewing them, and collecting their participating signatures and samples. Adults were selected by a sampling frame for survey research created by Walter Díaz from the UPRM Center for Applied Social Research. The work was supported by grants from the National Science Foundation and UPRM Seed Money programs of the College of Arts and Sciences and the Center for Research and Development.

## Notes

1. These restriction sites and motifs were -29 *Dde*I, +951 *Dpn*II, -3388 *Dde*I, +5054 *Rsa*I, +7750 *Dde*I, +7859 *Dpn*II, -8565 *Dpn*II, +9052 *Hae*II, -9253 *Hae*III, -9504 *Alu*I, -11793 *Dde*I, -11924 *Taq*I, -13633 *Hae*III, +16049 *Rsa*I, -16145 *Hae*III. They were sufficient to determine RFLP haplotype identity because all samples had already been tested for the following sites: 7013 *Rsa*I, 10394 *Dde*I, 10397 *Alu*I, 13259 *Alu*I, 16517 *Hae*III.

# 4

# Pre-Columbian Archaeology of Cuba

## A Study of Site Distribution Patterns and Radiocarbon Chronologies

JAGO COOPER

## Abstract

This chapter collates archaeological data from Cuba that can be used to identify and analyze site distribution and site chronologies. The study uses extant data from previous attempts to create a national sites and monuments record for Cuba and includes new information from recent archaeological projects to provide the names and locations of 1,061 archaeological sites in Cuba, details of artifact assemblages recovered from each site, and current site classifications based on existing archaeological frameworks. Here I discuss how spatial patterns in this data can be observed using geographical information systems (GIS) applications. In addition, all known radiocarbon determinations from archaeological sites in Cuba were collated. These 140 radiocarbon determinations are reviewed and the archaeological meaning of their calibrated dates discussed. The creation of a comprehensive database of Cuban archaeology, integrated with a GIS platform for data projection and analysis, provides a useful framework for studying archaeology in Cuba.

## Resumen

Este capítulo intenta reunir los datos arqueológicos acerca de Cuba que podrían ser utilizados para identificar y analizar la distribución de sitios y su cronología. Esta investigación está construída sobre valiosos intentos anteriores de crear un registro nacional de sitios y monumentos arqueológicos de Cuba. Este estudio proporciona los nombres y localizaciones de 1,061 sitios arqueológicos conocidos, detalles de los conjuntos de artefactos recolectados en cada sitio y la vigente clasificación de sitios desde una particular perspectiva arqueológica. Este capítulo incluye una discusión acerca de como los patrones espaciales pueden ser observados usando aplicaciones de un sistema de información geográfica (SIG). Además, todos los fechados radiocarbónicos de sitios arqueológicos en Cuba fueron reunidos. Estos 140 fechados son revisados y se discuten los significados arqueológicos de sus calibraciones y contexto. La creación de una completa base datos de la arqueología Cubana, integrada en una plataforma de SIG para visualizar y analizar la información, proporciona un sistema útil para estudiar el patrimonio cultural en Cuba.

## Introduction

Cuba has an important contribution to make to this book on the study of pre-Columbian settlement of the Caribbean. Geographically, the nation of Cuba is an island archipelago that constitutes 47 percent of the land area of the Caribbean and archaeologically has some of the earliest evidence of human colonization in the Caribbean. Sites such as Levisa and Canimar Abajo have been identified as providing some of the earliest evidence, primarily from lithic artifacts, for human settlement in the Caribbean (Martínez Fuentes et al. 2003: 64; Wilson et al. 1998). A wealth of archaeological information has been generated by over one hundred years of research in Cuba (Dacal Moure 2006; Marichal García 1995; Nuñez Jiménez 1992: 16). However, much of this information is not always easily accessible within Cuba or well disseminated internationally.

In this chapter, I discuss the collation of existing archaeological data from Cuba into a national database. The term *national* is used to define the scale of study that includes available information from the country of Cuba. This database was then used to study spatial and temporal evidence for pre-Columbian archaeology. Site locations were identified and projected to facilitate spatial analysis. Radiocarbon determinations were collated and calibrated to assess the temporality of human settlements in Cuba. Methodological issues of dealing with this macro-scale of archaeological data are then evaluated in light of the sample of archaeological information available.

## National Sites and Monuments Record

There have been a number of previous attempts by archaeologists to collate archaeological data from Cuba into a centralized computer-based system. A recent discussion of plans to create an atlas of Cuban heritage has been published in the Cuban journal *Catauro* (Departamento de Arqueología de Centro de Antropología 2003: 199), but the results of this project have not yet been published. The most recent published version of a national database was in 1995 (Febles Duenas and Martínez 1995). A CD-ROM was produced of archaeological census data from 975 archaeological sites, which built upon earlier attempts by Febles Duenas and colleagues (1987) and Rives Pantoja and colleagues (1991) to computerize archaeological data in Cuba.

Although the 1995 census is now over ten years old, it still provides the most complete summary of archaeological site data for Cuba. The census data set includes categories of site information recorded on predefined document templates and can be searched for information about individual sites, but there is no means of analyzing the data through relational queries. The majority of the sites have map coordinates, but the maps to which they refer are not easily available; as a result, the actual locations of many of the sites are not widely known. Establishing the locations of archaeological sites within a nationwide framework is necessary before studying site distribution patterns becomes possible.

In addition to the 1995 census, there is also a substantial body of data in the archaeological literature (Dacal Moure 2006). This includes information on new archaeological sites excavated since 1995, as well as supplementary information on existing sites. By extracting the data from the 1995 census and adding to it data from the available literature, I was able to create a relational database of 1,061 archaeological sites in Cuba.

## Database Design and Data Organization

The relational database was designed with forty-one related tables for data entry of available archaeological, geographic, and environmental evidence. The categories of information for each site are, to a large extent, reliant on the nature of the existing data. This is a limitation of the database, as the level of detail of available information for each site varies and the basis on which previous conclusions have been made by archaeologists is not always clear or well referenced. Therefore, a number of the categories of information used are based on the preexisting categories recorded during the 1995 census to provide a standardized framework that enables intersite comparisons. The primary site table includes an individual site reference number, the site name, projected coordinates, the elevation, the province, the municipality, topography, soil, geology, site artifact categories, and individual artifact details, as well as classifications of site subsistence, site economy, site phase, and site chronology.

## Site Locations

A map of Cuba was generated using the global shoreline data available from the National Geophysical Data Center of the National Oceanic and Atmospheric Administration. These data were projected in Universal Transverse Mercator (UTM) World Geodetic System (WGS) 84. This projection system was selected based on its worldwide popularity and compatibility with existing global positioning systems (GPS) and provides a good template for mapping locations of archaeological sites in Cuba. Two methods were used to identify and project the archaeological site locations: geographic coordinate reprojection and site point digitization.

For sites with existing coordinates (either map coordinates or latitude and longitude coordinates), it was possible to reproject them into UTM WGS 84. Experimental reprojections were tested using sites with both recorded map coordinates and known locations in UTM WGS 84 that were recorded during recent archaeological fieldwork (Cooper et al. 2006; Valcárcel Rojas et al. 2006). My study revealed that the different maps all used the national grid-projected coordinate system of either North American Datum (NAD) 1927 Cuba Norte or NAD 1927 Cuba Sur. By cross-referencing the site location with the known province and municipality of each site in the database, it was possible to identify the projection system used for each site and reproject all of the sites to UTM WGS 84.

The second method used to identify site location was the production of high-resolution scanned images of existing maps with archaeological site locations. These scanned images were georeferenced to the existing map of Cuba. The archaeological sites could then be manually digitized to provide point data with x-y coordinates in UTM WGS 84. The accuracy of these site locations is dependent on the quality of the original site maps; confidence levels in the accuracy of site locations were recorded in the database.

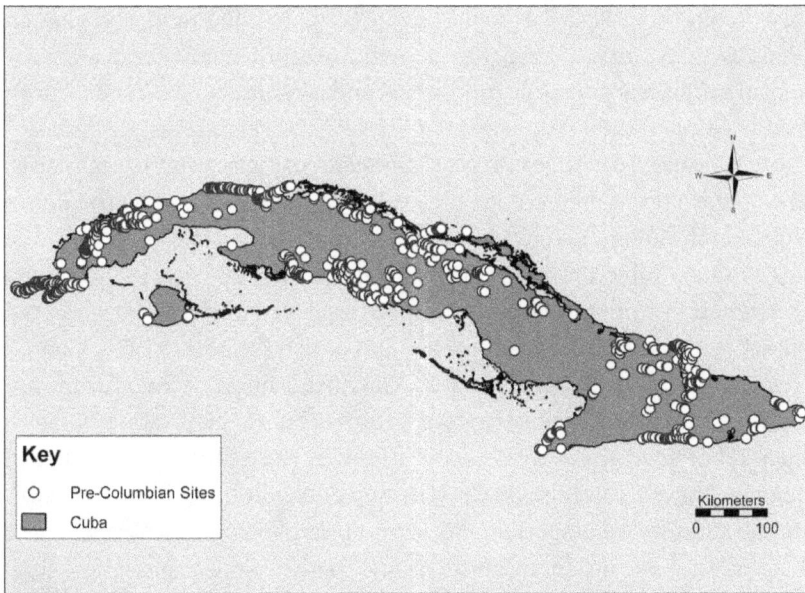

Figure 4.1. Distribution of archaeological sites in Cuba.

The methods described allowed the locations of 998 archaeological sites in Cuba to be identified and reprojected in order to study site distribution patterns. Figure 4.1 shows the site distribution of these 998 archaeological sites in Cuba.

## Site Classification

Frameworks for the classification of archaeological sites in Cuba reflect the influence of theoretical and methodological approaches that have emerged in Cuban archaeology. It is necessary to use existing classificatory frameworks for sites in order to investigate existing archaeological data at a national scale. There has been much debate about the suitability of different systems of site classification within Cuban archaeology. Discussion of these frameworks and the context of their development can be found in numerous publications (Berman et al. 2005; Dacal Moure and Watters 2005; Davis 1996; Godo 1997; Hernández Oliva and Arrazcaeta Delgado 2004; La Rosa Corzo 2003; Marichal García 1995; Torres Etayo 2004; Trincado Fontán 2000). Two major systems of site classification that have been used extensively in Cuba since the 1960s provide a standard framework for a large number of sites. The first classification framework, promoted by Tabio and

Rey (Tabio 1974, 1984; Tabio and Rey 1979; Tabio 1995; Tabio and Guarch 1966), is, in its simplest form, based on the presence or absence of archaeological evidence for ceramic production and agriculture. This classification has three categories: (1) *preagroalfarero* (preagroceramicist), (2) *protoagricola* (protoagriculturalist) and (3) *agroalfarero* (agricultural-ceramicist). The Spanish terms are part of a theoretical framework that is particular to Cuban archaeology, and to avoid confusion, the Spanish terms are used in this chapter (Tabio 1984). The second classification framework, promoted by Guarch Delmonte, adopts a more focused, economic approach based on artifact assemblages from each site (Guarch Delmonte 1990; Guarch Delmonte et al. 1995). Artifacts were classified using an economic framework as evidence of subsistence appropriation, or production. Guarch then subdivided these two classifications of site economy into phases, namely, Phase 1, hunting; Phase 2, fishing and collecting; Phase 3, incipient agriculture (all associated with appropriative economies); and Phase 4, agriculture for productive economies. These economic phases are then further categorized into cultural variants based on site and regional variations in material culture: Phase 1, the Seboruco culture; Phase 2, including the Guanahacabibes and Guacanayabo cultures; Phase 3, including Canimar and Arroyo del Palo; and Phase 4, including Damajayabo, Bayamo, Cunagua, Baní, and Maisí cultural variations.

Both systems of site classification have been discussed and critiqued during recent debates within Cuban archaeology (Godo 1997; La Rosa Corzo 2003; Torres Etayo 2004). However, these two existing classification systems are currently the only frameworks that provide a nationwide perspective on site classifications.

The spatial projections of sites based on the two site classification methods are illustrated in figures 4.2 and 4.3. Figure 4.2 includes site classifications for 983 sites and indicates the absence of agroalfarero sites in the west of Cuba. It also reveals a widespread distribution of preagroalfarero sites throughout the country with a concentration of sites in the western province of Pinar del Rio. Discussion of this preagroalfarero concentration in the west of Cuba and the association with ethnohistorical references to the Guanahatabey or Guanahacabibes has sparked debate over recent years (Keegan 1994: 271; Keegan 1989). Another popular hypothesis in Cuban archaeology is that intensive agricultural societies with elaborate artistic traditions spread from eastern Cuba westward (Guarch Delmonte 1978; Valcárcel Rojas 2002), influenced by their interaction with the societies on

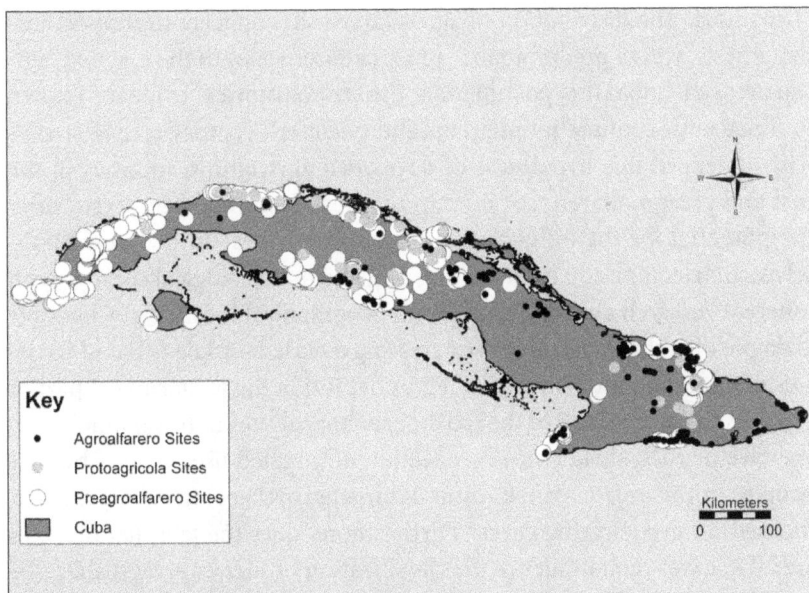

Figure 4.2. Distribution of sites based on the classification framework developed by Ernesto Tabio.

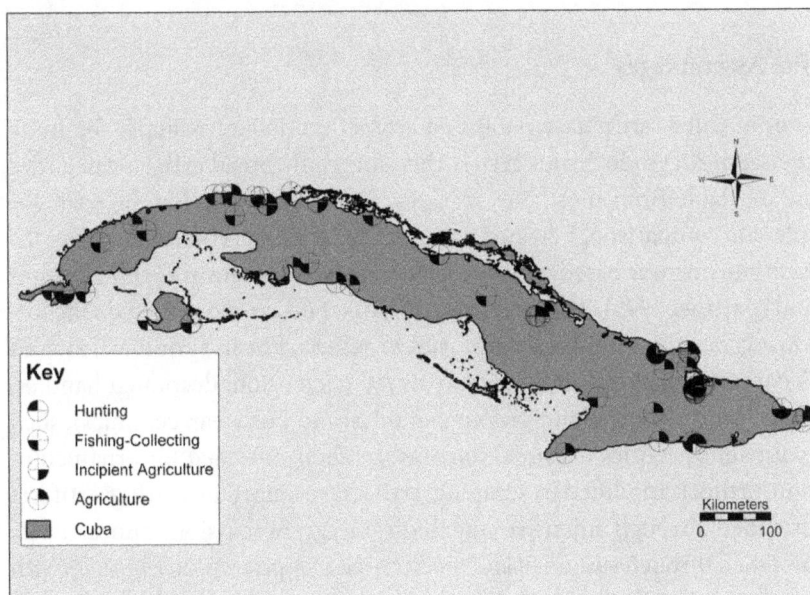

Figure 4.3. Distribution of sites based on the classification framework developed by José Guarch Delmonte.

Hispaniola. The distribution of agroalfarero sites appears to support this hypothesis, with a predominance of agroalfarero sites in the east and central areas of Cuba. It is possible that the ethnohistorical evidence known to nineteenth-century antiquarians and twentieth-century archaeologists has influenced this hypothesis of developed agricultural societies in the east and preagroalfarero societies in the west. It is important to consider the potential for the influence of preconceived ideas to manifest themselves archaeologically by attracting targeted archaeological surveys with inherent research agendas to different geographical locations. A possible example of archaeological survey creating a biased sample is found in the western Sandino municipality of Pinar del Rio in the westernmost part of Cuba. Is the fact that 90 of the 103 archaeological sites in this municipality are cave or rock-shelter sites a reflection of targeted use of caves by past peoples in this region or the result of targeted archaeological survey that focused on investigating caves? Furthermore, does the fact that the sites are all in caves then influence the classification of sites as preagroalfarero? In order to further examine these questions, it is possible to investigate the spatial distribution of different categories of material culture at archaeological sites rather than rely on existing site classifications alone.

## Site Assemblages

Sites in Cuba rarely have published artifact catalogues available for study (Febles 1982; Godo Torres 1994). Therefore, only broad artifact categories are available from most sites in Cuba. In order to provide the basis for intersite comparison, I created a standardized list of artifact categories for each site that was based on those used in the Febles census (Febles Duenas and Martínez 1995). The categories of artifact descriptions include the following: ceramics (with subcategories of vessels, *burens* [griddles], incised decoration, appliqué decoration, painted decoration, decorated handles, European-influenced indigenous ceramics, and European ceramics); shell (with subcategories of faunal remains, artifacts modified for ornamentation, artifacts modified by scraping, artifacts modified by cutting, artifacts modified through intensive and high-energy percussion, and artifacts modified through sustained and medium-energy percussion); burials (with subcategories of primary burials, secondary burials, and burials with grave goods); bone (with subcategories of faunal remains, bone modified by cutting, bone modified by scraping, and bone modified for ornamentation);

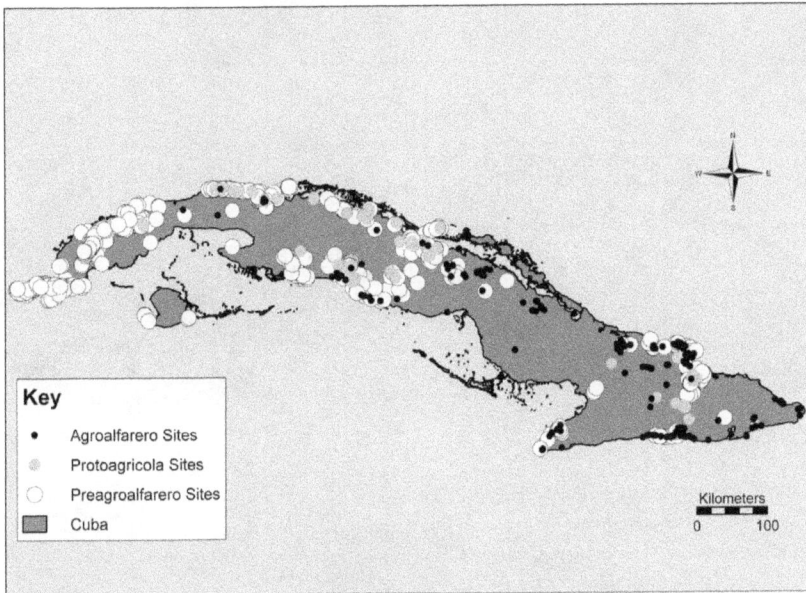

Figure 4.4. Locations of excavated burials.

wood (wood modified for ornamentation, worked wood, and unworked wood); paints and dye materials; metals (colonial and nonlocal metal, European metal, and nonferrous metal); stone (stone modified for ornamentation, stone modified by hammering, stone modified by polishing, lithics modified by knapping, and unmodified); and textiles.

The spatial distribution of the sites with each of these categories of artifact was then projected. Patterns in the distribution of European-influenced material culture at indigenous sites provide an interesting topic of research, but this is not discussed here. Human remains are recorded at 176 sites throughout Cuba (see figure 4.4). There is a widespread distribution of burials with associated grave goods that includes sites classified as preagroalfarero in the west and agroalfarero in the east. Shell and stone artifacts are the most common artifact categories, found at over 90 percent of archaeological sites in Cuba, and they have a relatively uniform distribution throughout the island. There are only a limited number of sites where wood and textiles have been recovered, and spatial patterns in distribution appear to reflect local environmental conditions rather than any archaeologically significant pattern.

The spatial distribution of sites with indigenous ceramics is illustrated

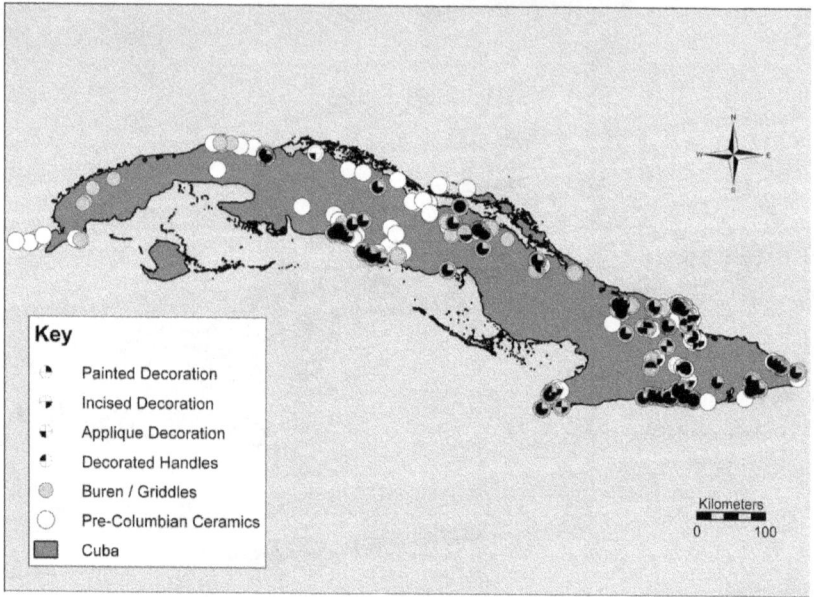

Figure 4.5. Distribution of pre-Columbian ceramics.

in figure 4.5. This map appears to reflect a broad pattern of ceramic distribution similar to sites classified as agroalfarero, but it also shows a subtler pattern in ceramic style distribution. There appears to be a concentration of elaborate decoration styles in central and eastern Cuba. There is evidence of buren fragments found in the western province of Pinar del Rio at the rock-shelter site of Solapa de Nora and four cave sites, including Cueva del Chino, Cueva de Evaristo, Cueva de la Bibijagua, and Cueva de la Pintura. There are also vessel fragments found in seven sites in the most western municipality of Sandino in Pinar del Rio, including Cueva de Paulino, Cueva de Bolondron, Cueva de la Viuda, Cueva del Resguardo, Cueva del Negro, Cueva de la Pintura, and Cueva del Agua. These sites are all classified as preagroalfarero, or appropriative fisher-collectors. This highlights how artifact distribution patterns can complement the existing site classifications and provide a more detailed framework for interpreting pre-Columbian settlement in Cuba. However, spatial patterns in artifact and site distribution need to be complemented by an understanding of temporal context for past human activity that can only be gained through a better understanding of site chronology.

## Site Chronologies

In Cuba, artifact typologies only provide extended, relative chronological ranges. Consequently, sites and archaeological contexts are normally allocated broad dates based on the presence or absence of diagnostic artifacts such as worked lithics, ground stone tools, shell artifacts, and ceramics. Guarch Delmonte and colleagues (1995) allocated chronological ranges to site phases based on the presence or absence of these artifacts and the existing radiocarbon laboratory dates for certain type sites. These chronological ranges include hunters (6000–2500 BC), fisher-collectors (2500 BC–AD 1500), incipient agriculturalists (400 BC–AD 1500), and agriculturalists (AD 600–1500). Such broad periods reflect a lack of well-defined and securely dated artifact typologies in Cuban archaeology.

## Radiocarbon Chronologies

Collating existing radiocarbon determinations from archaeological sites in Cuba is a useful starting point to begin framing a temporal context for pre-Columbian archaeology in Cuba. Radiocarbon determinations can provide a useful method for establishing relative and absolute site chronologies. However, comparisons of radiocarbon determinations are fraught with methodological issues that can limit the usefulness of direct association between radiocarbon dates and archaeological context, as well as comparison between radiocarbon dates. Radiocarbon determinations from the site of Vega de Palmar in Cuba are listed in the first volume of the journal *Radiocarbon* in 1959 (Deevey et al. 1959), illustrating the fact that radiocarbon dating has a long history of use in Cuban archaeology. During this period, few radiocarbon determinations appear to have been calibrated (Ulloa Hung and Valcárcel Rojas 2002; Wilson et al. 1998), and laboratory dates are often misrepresented as being calendrically significant, having been cited from secondary literary sources. This repetition of laboratory dates is not always explicit, and the chronological significance of a given date is often unclear. In addition, many of the radiocarbon dates in the literature are listed without the necessary information required to interpret the archaeological significance of the date, such as archaeological provenience, what material was dated, where and when it was dated, what (if any) calibration methods were used, and what error margins are involved. Without this important information, many of these dates cannot be used to provide a reliable indication of site chronology.

Attempts have been made recently to establish some standards for the use of radiocarbon dating in Caribbean archaeology (Fitzpatrick 2006). However, before this can be done in Cuba, all of the available information on radiocarbon determinations in Cuba needs to be collated. Therefore, I reviewed the extant literature for details of radiocarbon determinations from Cuba. Alternative sources of data were cross-referenced to create a list of 140 radiocarbon determinations from archaeological sites in Cuba, illustrated in table 4.1 (Deevey et al. 1959; Godo Torres 1994; Jardines Macias and Calvera Roses 1999; Jouravleva and González 2000; Kozlowski 1974; Martínez Fuentes et al. 2003; Mielke and Long 1969; Navarrete Pujol 1990; Pazdur et al. 1982; Pendergast et al. 2002; Pino 1995; Rankin Santander 1994; Stuckenrath and Mielke 1973; Trincado Fontán and Ulloa Hung 1996; Ulloa Hung and Valcárcel Rojas 2002; Valcárcel Rojas 2002; Vinogradov et al. 1968; Wilson et al. 1998).

All radiocarbon dates represent a statistical calculation with inherent margins of error, and archaeologists must bear the responsibility for assessing how the potential errors of each radiocarbon date affect their usefulness in archaeological interpretation. Therefore, a summary of radiocarbon determinations from archaeological sites in Cuba provides useful information with which to evaluate potential margins of error for each radiocarbon determination.

## Radiocarbon Chronologies Discussion

As is shown in table 4.1, 135 of the 140 known radiocarbon determinations from Cuba have the necessary stratigraphic information to facilitate their use in identifying site chronologies. A number of the radiocarbon dating laboratories used to date archaeological samples from Cuba, such as Gliwice (GD-624), Leningrad (LE-4290), and Vernadsky (MO-399), were in the former Soviet Union, and the history and methods of these laboratories are not widely known in Europe and North America (Taylor 1987: 168). Publications on radiocarbon determinations from these laboratories appear to indicate that reliable methods were used and that the laboratories were part of the interlaboratory cross-checks carried out between radiocarbon laboratories to verify international standards that began in the 1960s (Pazdur et al. 1982; Vinogradov et al. 1968).

The use of radiocarbon dates from the early 1950s and 1960s raises some questions about their acceptability. Sample Y-465, from Vega del Palmar,

Table 4.1. Radiocarbon Dates from Archaeological Sites in Cuba

| Site name | Laboratory code | Lab. date BP | ± | Stratigraphic context |
|---|---|---|---|---|
| Abra del Cacoyuguin I | Beta-133948 | 1640 | 130 | Excavation 1, enlargement 1, level 30–40 cm |
| Abra del Cacoyuguin I | Beta-133947 | 1210 | 60 | Excavation 1, enlargement 1, level 10–20 cm |
| Abra Rio Cacoyuguin II | Beta-133950 | 2780 | 40 | Excavation 2, grid square 1, level 40–50 cm |
| Abra Rio Cacoyuguin II | Beta-133951 | 3720 | 70 | Excavation 2, grid square 1, level 50–60 cm |
| Abra Rio Cacoyuguin IV | Beta-140079 | 4180 | 80 | Cut 1, level 30–40 cm |
| Aguas Gordas | GD-1054 | 485 | 50 | Mound 2, pit 1, level 75–100 cm |
| Aguas Gordas | GD-621 | 705 | 65 | Midden 2, pit 1, level 1.25–1.5 m; assoc. with ceramics, shell and stone artifacts |
| Aguas Gordas | GD-620 | 165 | 60 | Midden 2, pit 1, level 50–75 cm; assoc. with ceramics and some shell and stone artifacts |
| Aguas Gordas | GD-1055 | 575 | 60 | Midden 2, pit 1, level 1–1.25 m |
| Aguas Gordas | MO-399 | 1000 | 105 | Midden 1, sample depth 1.75 m |
| Arroyo del Palo (Mayari) | Y-1556 | 970 | 80 | Cave 1, sample depth 25 cm |
| Arroyo del Palo (Mayari) | Y-1555 | 760 | 60 | Trench 2b, level 75–100 cm, sample depth 75 cm |
| Belleza | Unknown-4 | 1120 | 60 | Trench 1, level 40 cm |
| Birama | Unknown-5 | 820 | 40 | No information |
| Cabagan | Unknown-6 | 1080 | 20 | No information |
| Caimanes III | UM-1953 | 1745 | 175 | Test pit 4, sample depth 38 cm |
| Canimar 1 | GD-203 | 1010 | 110 | Sample depth 70–80 cm; unsecure stratigraphy |
| Canimar Abajo | UBAR-170 | 4270 | 70 | Sample depth between 30 and 60 cm |
| Canimar Abajo | UBAR-171 | 4700 | 70 | Sample depth 1.65 m |
| Catunda | Beta-93862 | 1890 | 60 | Trench 2, level 40 cm |
| Catunda | Beta-93866 | 1850 | 50 | Trench 1, level 30 cm |
| Catunda | Beta-140078 | 1280 | 60 | Trench 5, level 20–30 cm |
| Chorro de Maíta | Beta-148955 | 360 | 80 | Skeleton 39, depth 79 cm |
| Chorro de Maíta | Beta-148957 | 730 | 60 | Unit 5, grid square 2, natural layer 1, spit depth 30–50 cm |
| Chorro de Maíta | Beta-148956 | 870 | 70 | Skeleton 25, depth 88 cm |

continued

Table 4.1—*Continued*

| Site name | Laboratory code | Lab. date BP | ± | Stratigraphic context |
|---|---|---|---|---|
| Corinthia III | Beta-133953 | 2220 | 70 | Excavation 3, encaque 3, level 10–20 cm |
| Corinthia III | Beta-133952 | 2300 | 60 | Excavation 4, encaque 2, layer 1 |
| Corinthia III | Beta-140080 | 1700 | 70 | Unit III, level 0–10 cm |
| Cueva 1, Punta del Este | GD-618 | 910 | 85 | Block I, sec. A, level 50–75 cm, sample depth 57 cm; assoc. with shell and stone artifacts |
| Cueva 4, Punta del Este | LC-H 1106 | 1100 | 130 | Test Pit, 1×0.5 m, sample depth 38 cm |
| Cueva de la Lechuza | LE-4281 | 2610 | 120 | Test pit 1, block 1, level 2.15 m |
| Cueva de la Lechuza | LE-4290 | 2610 | 120 | Test pit 1, block 1, level 2.05 m |
| Cueva de la Lechuza | LE-4283 | 5270 | 120 | Test pit 1, block 1, level 1.95 m |
| Cueva de la Lechuza | LE-4269 | 1470 | 110 | Test pit 1, block 1, level 25 cm |
| Cueva de la Lechuza | LE-4287 | 3030 | 180 | Test pit 1, block 1, level 1.65 m |
| Cueva de la Lechuza | LE-4275 | 2580 | 90 | Test pit 1, block 1, level 2.35 m |
| Cueva de la Lechuza | LE-4288 | 3030 | 180 | Test pit 1, block 1, level 1.55 m |
| Cueva de la Lechuza | LE-4271 | 2380 | 80 | Test pit 1, block 1, level 75 cm |
| Cueva de la Lechuza | LE-4272 | 2750 | 160 | Test pit 1, block 1, level 65 cm |
| Cueva de la Lechuza | LE-4267 | 2220 | 160 | Test pit 1, block 1, level 35 cm |
| Cueva de la Lechuza | LE-4274 | 2030 | 160 | Test pit 1, block 1, level 45 cm |
| Cueva de la Lechuza | LE-4282 | 2930 | 300 | Test pit 1, block 1, level 1.25 m |
| Cueva de la Lechuza | LE-4276 | 2250 | 150 | Test pit 1, block 1, level 55 cm |
| Cueva de la Lechuza | LE-4270 | 3110 | 180 | Test pit 1, block 1, level 1.05 m |
| Cueva de la Lechuza | LE-4273 | 2420 | 100 | Test pit 1, block 1, level 95 cm |
| Cueva de la Lechuza | LE-4279 | 2390 | 170 | Test pit 1, block 1, level 85 cm |
| Cueva de la Pintura | GD-1046 | 2840 | 60 | Excavation unit 2, block 5, sec. D, level 1.25–1.5 m; assoc. with shell and stone artifacts |
| Cueva de la Pintura | GD-613 | 2880 | 70 | Excavation unit 2, block 5, sec. D, level 1.5–1.75 m; assoc. with shell and stone artifacts |
| Cueva de la Pintura | GD-591 | 2930 | 80 | Excavation unit 1, block 1-i, sec. D, level 1.5–1.8 m; assoc. with shell and stone artifacts |

| Site | Lab number | Date | ± | Provenience |
|---|---|---|---|---|
| Cueva de la Pintura | GD-1039 | 2160 | 55 | Excavation unit 1, block 1-i, sec. A, level 50–75 cm; assoc. with shell and stone artifacts |
| Cueva de la Pintura | GD-614 | 2720 | 65 | Excavation unit 2, block 5, sec. D, level 1–1.25 m; assoc. with shell and stone artefacts |
| Cueva de la Pintura | GD-601 | 2805 | 60 | Excavation unit 1, block 1-i, sec. D, level 1–1.25 m; assoc. with shell and stone artefacts |
| Cueva del Perico I | GD-616 | 1350 | 70 | Trench 2, sec. 2, level 1.5–1.75 m; assoc. with human burials and shell and stone artifacts |
| Cueva del Perico I | GD-1051 | 1990 | 80 | Trench 1, sec. 1, level 1.3–1.4 m |
| Cueva del Perico I | GD-617 | 1495 | 60 | Trench 1, sec. 1, level 1–1.2 m; assoc. with human burials and shell and stone artifacts |
| Cueva Funche | SI-426 | 2070 | 150 | Block II, sec. A, sample depth 50 cm |
| Cueva Funche | SI-429 | 4000 | 150 | Block III, sec. A, sample depth 1.72 m |
| Cueva Funche | SI-428 | 3110 | 200 | Block III, sec. A, sample depth 1.40 m |
| Cueva Funche | SI-427 | 2510 | 200 | Block II, sec. D, sample depth 55 cm |
| Damayajabo | Y-1764 | 3250 | 100 | Trench 51, level 134 cm |
| Damayajabo | Y-1994 | 1120 | 160 | Sample found in association with ceramics |
| El Boniato (El Palmar) | Beta-148958 | 670 | 70 | Unit 2, grid square 9, natural layer 2, spit depth 40–50 cm |
| El Convento | GD-1053 | 665 | 50 | Pit 2, level 25–50 cm, sample depth 45 cm |
| El Convento | Unknown-7 | 400 | 20 | No information |
| El Guafe I | FS AC 2419 | 690 | 50 | Block 2, natural layer 2, sample depth 30 cm |
| El Guafe I | FS AC 2420 | 450 | 35 | Block 1, sec. 2 & 4, layer 3, sample depth 50 cm |
| El Morrillo | SI-353 | 590 | 90 | Block 9-q, sec. B, sample depth 45 cm; assoc. with ceramics, shell and stone artifacts |
| El Paraiso | Unknown-8 | 1130 | 150 | Test pit 1, 1×1 m, level 20–30 cm |
| El Porvenir | Beta-148960 | 500 | 50 | Unit 5, grid square b, natural layer 1, spit depth 40–50 cm |
| El Purial | UBAR-169 | 3060 | 180 | Level 40 cm (approx.) |
| Esterito | SI-350 | 500 | 100 | Midden 1, trench 1, sec. D, sample depth 1.15 m; assoc. with ceramics, shell and stone artifacts |

continued

Table 4.1—Continued

| Site name | Laboratory code | Lab. date BP | ± | Stratigraphic context |
|---|---|---|---|---|
| Esterito | SI-349 | 550 | 150 | Midden 1, trench 1, sec. C, sample depth 45 cm; assoc. with ceramics, shell and stone artifacts |
| Herradura 1 | Beta-140075 | 2050 | 70 | Cut 5, level 0–10 cm |
| Jorajuria | LE-1783 | 4110 | 50 | Pit 1, 1×1 m, level 80–90 cm |
| Jorajuria | LE-1784 | 3870 | 40 | Pit 1, 1×1 m, level 40–50 cm |
| Jorajuria | LE-1782 | 3760 | 40 | Pit 1, 1×1 m, level 60–70 cm |
| Jucaro | Beta-148949 | 690 | 60 | Cut A, natural layer 1, spit depth 20–40 cm |
| La Escondida de Bucuey | Unknown-9 | 1060 | 150 | Test pits 3 & 4, 1×1 m; level 2–3 m |
| La Guira | Beta-140077 | 1390 | 70 | Trench 1, level 19 cm |
| La Guira de Barajagua | SI-351 | 590 | 100 | Midden 1, trench 1, sec. B, sample depth 90 cm; assoc. with ceramics, shell and stone artifacts |
| La Luz | Beta-93863 | 1350 | 50 | Test excavation 3, level 1.20 m |
| Laguna de Limones | SI-348 | 640 | 120 | Midden 2, trench 2, sec. D, sample depth 40 cm |
| Levisa 1 (Far. de Lev.) | MC-860 | 4420 | 100 | Sec. i-i, level 55–60 cm, layer 6 |
| Levisa 1 (Far. de Lev.) | GD-250 | 5140 | 170 | Sec. i-i, 85–90 cm |
| Levisa 1 (Far. de Lev.) | MC-859 | 4240 | 100 | Sec. i-i, level 55–60 cm, layer 6 |
| Levisa 1 (Far. de Lev.) | GD-204 | 3460 | 160 | Sec. i-i, layer v, 50–55 cm |
| Levisa 8 (Cueva S. Rita) | LE-2720 | 2680 | 40 | Unit 3, sec. 23 a, 40–50 cm, layer 1 |
| Levisa 8 (Cueva S. Rita) | LE-2718 | 2610 | 40 | Unit 3, sec. 45, 20–22 cm, layer 1 |
| Levisa 8 (Cueva S. Rita) | LE-2719 | 2160 | 40 | Unit 2, sec. 25, 20–40 cm, layer 3 |
| Levisa 8 (Cueva S. Rita) | LE-2717 | 2010 | 40 | Unit 3, sec. 35 a, 20–30 cm, layer 2/3 |
| Loma de la Campana | GD-1057 | 490 | 45 | Midden 2, block i, sec. C, level 50–75 cm; assoc. with ceramics, shell and stone artifacts |
| Loma de la Campana | GD-624 | 505 | 40 | Midden 2, block ii, sec. D, level 75–100 cm; assoc. with ceramics, shell and stone artifacts |
| Loma de la Campana | GD-1056 | 600 | 55 | Midden 2, block ii, sec. D, level 1–1.50 m; assoc. with ceramics, shell and stone artifacts |

| | | 970 | 100 | |
|---|---|---|---|---|
| Loma de la Forestal | SI-352 | 970 | 100 | Midden 9, trench 1, sec. A, sample depth 70 cm; assoc. with ceramics, shell and stone artifacts |
| Loma de Ochile | FS AC 2414 | 770 | 35 | Block 2, sec. 3, natural layer 1, sample depth 10–30 cm |
| Loma de Ochile | FS AC 2415 | 690 | 50 | Block 2, sec. 1, 2 & 3, natural layer 2, sample depth 30–40 cm |
| Loma de Ochile | FS AC 2416 | 660 | 35 | Block 1, sec. 1–2, natural layer 2, sample depth 30–60 cm |
| Loma de Ochile | FS AC 2417 | 620 | 30 | Block 1, sec. 2, natural layer 3, sample depth 60–80 cm |
| Loma de Ochile | FS AC 2418 | 880 | 40 | Block 1, sec. 2, natural layer 4, sample depth 80–90 cm |
| Los Buchillones | TO-8070 | 280 | 60 | Post 4, structure f1-1 |
| Los Buchillones | TO-7627 | 460 | 50 | King post 1, structure d2-1 |
| Los Buchillones | TO-7628 | 560 | 50 | King post 2, structure d2-1 |
| Los Buchillones | TO-8067 | 240 | 60 | Post 1, structure f1-1 |
| Los Buchillones | TO-7624 | 1320 | 60 | Rafter 3, structure d2-1 |
| Los Buchillones | TO-7623 | 390 | 50 | Rafter 2, structure d2-1 |
| Los Buchillones | TO-7622 | 320 | 40 | Post 13, structure d2-1 |
| Los Buchillones | TO-7621 | 1404 | 60 | Post 12, structure d2-1 |
| Los Buchillones | TO-7620 | 430 | 50 | Post 7 sub, structure d2-1 |
| Los Buchillones | TO-7619 | 300 | 50 | Post 7, structure d2-1 |
| Los Buchillones | TO-7618 | 510 | 50 | Post 2, structure d2-1 |
| Los Buchillones | TO-8069 | 230 | 70 | Post 3, structure f1-1 |
| Los Buchillones | TO-8071 | 250 | 60 | Post 5, structure f1-1 |
| Los Buchillones | TO-8072 | 430 | 60 | Post 6, structure f1-1 |
| Los Buchillones | TO-7626 | 540 | 50 | Rafter 5, structure d2-1 |
| Los Buchillones | TO-8068 | 480 | 60 | Post 2, structure f1-1 |
| Los Buchillones | TO-7617 | 330 | 50 | Post 1, structure d2-1 |
| Los Buchillones | TO-7625 | 340 | 50 | Rafter 4, structure d2-1 |
| Los Chivos | Beta-140074 | 1150 | 60 | Trench 1, level 10–20 cm |
| Los Chivos | Beta-140076 | 2710 | 80 | Trench 1, level 45 cm |
| Los Pedregales | GD-619 | 1170 | 90 | Trench 2, sec. B, level 2–2.25 m, sample depth 2 m; assoc. with shell and stone artifacts |

continued

Table 4.1—*Continued*

| Site name | Laboratory code | Lab. date BP | ± | Stratigraphic context |
|---|---|---|---|---|
| Marien 2 | LV-2063 | 2020 | 80 | Excavation square m-07, level 20–30 cm |
| Marien 2 | LV-2062 | 780 | 100 | Excavation square ll-10, level 10–20 cm |
| Mejias | SI-347 | 1020 | 100 | Trench 1, sec. B, sample depth 45 cm |
| Mogote de la Cueva | Unknown-3 | 960 | 50 | No information |
| Mogote de la Cueva | SI-424 | 1620 | 150 | Trench 1, sample depth 35 cm; unsecure stratigraphy |
| Mogote de la Cueva | SI-425 | 650 | 200 | Trench 1, level 1, sample depth 1.25 m |
| Playita (Villa Clara) | Unknown-2 | 1280 | 20 | No information |
| Potrero del Mango | Y-206 | 810 | 80 | Midden 1, sec. y-5, level 75–100 cm |
| Potrero del Mango | Beta-148961 | 880 | 80 | Unit 1, grid square a, spit depth 80–90 cm |
| Potrero del Mango | Beta-148962 | 620 | 60 | Unit 2, grid square a, spit depth 1–1.1 m |
| Punta de Peque | Beta-93860 | 1400 | 60 | Trench 1, level 50 cm |
| San Benito | Beta-93851 | 2020 | 60 | Trench 2, level 40–50 cm |
| Vega del Palmar | Y-465 | 960 | 60 | Midden 150 cm deep, sample depth 105–120 cm; ceramics only found in the top two 15-cm spits |
| Ventas de Casanova | FS AC 2421 | 375 | 25 | Test trench, sec. 4, layer 1 & 2, sample depth 0–23 cm |
| Ventas De Casanova | FS AC 2424 | 475 | 35 | Block 1, sec. 1, layer 4, sample depth 60–80 cm |
| Ventas De Casanova | FS AC 2422 | 420 | 45 | Block 1, sec. 1 & 2, layer 3, sample depth 30–50 cm |
| Ventas De Casanova | FS AC 2423 | 315 | 45 | Block 1, sec. 1 & 2, layer 4, sample depth 50–60 cm |
| Victoria I | LC-H 565 | 960 | 50 | Block, sec. B, level 2–2.25 m |
| Victoria I | LC-H 1034 | 2070 | 110 | Block 1, sec. b, level 6.25–6.50 m |
| Victoria I | LC-H 1035 | 1450 | 70 | Block 1, sec. b, level 2–2.25 m |

was collected in 1956 and dated before 1959, for example, but whether the Libby half-life or the Cambridge half-life was used in its determination is unknown. This potential error can be accounted for by increasing the laboratory error for the laboratory date by 3 percent based on the difference between the two half-life calculations (Higham 2005). By 1970, the effects of isotopic fractionation on radiocarbon determinations were well known among the radiocarbon community, but they were considered to have been quite minor, and laboratories accounted for them by increasing the error margin by±80 years (Barker 1970: 39). By 1977, procedures for accounting for isotopic fractionation, based on the $\delta^{13}C$ of individual samples, were well established (Stuiver and Polach 1977: 356). However, it must be assumed that the radiocarbon determinations from before 1977 did not account for isotopic fractionation but merely increased the margins of error by an additional±80 years. Studies have indicated that isotopic fractionation can in fact lead to larger errors than originally anticipated when dating charcoal samples (Taylor 1987: 122). This must be taken into account when considering the use of pre-1977 radiocarbon determinations.

The archaeological context of the samples taken for radiocarbon dating, detailed in table 4.1, provides useful information for their interpretation. For example, the early laboratory date of sample LE-4283 from Cueva de la Lechuza does not appear to be corroborated by further dates taken from deeper stratigraphic levels at the site.

As discussed above, there remain a number of potential issues that may affect the direct comparison of radiocarbon dates. In particular, it is necessary to calibrate the laboratory dates in order to provide a more meaningful basis upon which to base discussion of site chronologies. Calibrated dates provide a more valid means of comparing radiocarbon determinations and also provide a more relevant chronology for comparisons with historical dates.

## Calibrated Radiocarbon Determinations

The laboratory dates were calibrated using OxCal 3.8. Samples from terrestrial sources were calibrated using IntCal04 (Reimer et al. 2004). Isotopic data for the bone samples were not available, and the potential for a marine diet of the inhabitants of El Chorro de Maíta must be considered when assessing the reliability of calibrated dates from samples Beta-148955 and Beta-148956 (Bayliss et al. 2004). The samples from marine sources

were calibrated using Marine04 (Hughen et al. 2004). Local marine reservoir effects have not been determined for Cuba, and regional data from
the Caribbean were considered but not used in this study (Reimer 2005;
Reimer et al. 2002). Further methodological issues surrounding the use
of marine shell should also be considered before using this sample type
as direct evidence of site chronology (Ascough et al. 2005a; Ascough et al.
2005b; Rick et al. 2005; Stuiver and Braziunas 1993). Calibrated dates for
all samples were calculated to 2σ (see table 4.2).

The calibrated data set allows identification of patterns in the radiocarbon chronologies. The calibrated marine shell dates of early ceramic
sites become more contemporaneous with the terrestrial charcoal dates
from similar sites discussed by Ulloa Hung and Valcárcel Rojas (Ulloa
Hung and Valcárcel Rojas 2002). Radiocarbon determinations from non-
archaeological contexts are not included in this summary of radiocarbon
dates from Cuba. However, calibrating the radiocarbon dates provides
a means for interdisciplinary comparison. Studies of the megafauna extinction chronology and their relationship with human colonization have
been an interesting area of research for many years (Koch and Barnosky
2006). One observation that can be made from the calibrated radiocarbon
dates from Cuba is that the latest known date for the survival of two sloth
species in Cuba comes from the sites of Cueva Beruvides (*Megalocnus
rodens;* 7270–6010 cal BP) and Las Breas de San Felipe (*Parocnus Brownii;*
6350–4950 cal BP) (Steadman et al. 2005: 11766). These dates are potentially contemporaneous with the earliest evidence for human colonization
taken from the archaeological sites of Cueva de la Lechuza (6298–5746 cal
BP), Levisa 1 (6288–5584 cal BP), and Canimar Abajo (5590–5300 cal BP).
Further research is required in order to generate a larger sample of dates
from both archaeological and paleozoological contexts before links can be
drawn between human settlement of Cuba and megafauna extinctions, but
this example illustrates the advantages of collating and calibrating radiocarbon determinations from archaeological contexts in Cuba.

One of the aims of this study is to analyze distribution patterns of sites
with calibrated radiocarbon chronologies. This can be accomplished by
mapping site locations from where the radiocarbon samples were collected. The samples were projected chronologically using the calibrated
date ranges to compare with settlement distribution. In order to provide
an illustrative example that is intelligible, the uppermost ranges of cali-

Table 4.2. Calibrated Date Ranges of Radiocarbon Determinations from Archaeological Sites in Cuba

| Site name | Laboratory code | Lab. date BP | Cal BP 2σ lower range | Cal BP 2σ upper range | Samples dated pre-1977 |
|---|---|---|---|---|---|
| Abra del Cacoyuguin I | Beta-133948 | Charcoal | 1866 | 1296 | |
| Abra del Cacoyuguin I | Beta-133947 | Charcoal | 1283 | 974 | |
| Abra Rio Cacoyuguin II | Beta-133950 | Charcoal | 2964 | 2779 | |
| Abra Rio Cacoyuguin II | Beta-133951 | Charcoal | 4256 | 3873 | |
| Abra Rio Cacoyuguin IV | Beta-140079 | Charcoal | 4867 | 4446 | |
| Aguas Gordas | GD-1054 | Charcoal | 624 | 480 | 1971 |
| Aguas Gordas | GD-621 | Charcoal | 734 | 550 | 1971 |
| Aguas Gordas | GD-620 | Charcoal | 307 | 1 | 1971 |
| Aguas Gordas | GD-1055 | Charcoal | 666 | 508 | 1971 |
| Aguas Gordas | MO-399 | Charcoal | 1149 | 692 | 1963 |
| Arroyo del Palo (Mayari) | Y-1556 | Charcoal | 1055 | 727 | 1965 |
| Arroyo del Palo (Mayari) | Y-1555 | Charcoal | 787 | 568 | 1965 |
| Belleza | Unknown-4 | Charcoal | 1176 | 927 | |
| Birama | Unknown-5 | Charcoal? | 793 | 674 | |
| Cabagan | Unknown-6 | Bone | 1054 | 934 | |
| Caimanes III | UM-1953 | Charcoal | 2060 | 1300 | |
| Canimar 1 | GD-203 | Charcoal | 1174 | 692 | 1973 |
| Canimar Abajo | UBAR-170 | Charcoal | 5030 | 4622 | |
| Canimar Abajo | UBAR-171 | Charcoal | 5590 | 5300 | |
| Catunda | BETA-93862 | Charcoal | 1950 | 1700 | |
| Catunda | BETA-93866 | Charcoal | 1894 | 1631 | |
| Catunda | BETA-140078 | Charcoal | 1302 | 1062 | |
| Chorro de Maíta | BETA-148955 | Human bone | 533 | 154 | |
| Chorro de Maíta | BETA-148957 | Charcoal | 740 | 561 | |
| Chorro de Maíta | Beta-148956 | Human bone | 930 | 673 | |
| Corinthia III | Beta-133953 | Marine shell | 1986 | 1650 | |
| Corinthia III | Beta-133952 | Marine shell | 2078 | 1770 | |
| Corinthia III | Beta-140080 | Marine shell | 1380 | 1114 | |
| Cueva 1, Punta del Este | GD-618 | Charcoal | 969 | 675 | 1967 |
| Cueva 4, Punta del Este | LC-H 1106 | Charcoal | 1292 | 735 | |
| Cueva de la Lechuza | LE-4281 | Charcoal | 2958 | 2352 | |
| Cueva de la Lechuza | LE-4290 | Charcoal | 2958 | 2352 | |
| Cueva de la Lechuza | LE-4283 | Charcoal | 6298 | 5746 | |
| Cueva de la Lechuza | LE-4269 | Charcoal | 1568 | 1178 | |
| Cueva de la Lechuza | LE-4287 | Charcoal | 3638 | 2762 | |
| Cueva de la Lechuza | LE-4275 | Charcoal | 2856 | 2358 | |
| Cueva de la Lechuza | LE-4288 | Charcoal | 3638 | 2762 | |
| Cueva de la Lechuza | LE-4271 | Charcoal | 2720 | 2181 | |
| Cueva de la Lechuza | LE-4272 | Charcoal | 3328 | 2460 | |
| Cueva de la Lechuza | LE-4267 | Charcoal | 2719 | 1864 | |
| Cueva de la Lechuza | LE-4274 | Charcoal | 2349 | 1610 | |
| Cueva de la Lechuza | LE-4282 | Charcoal | 3834 | 2346 | |

*continued*

Table 4.2—*Continued*

| Site name | Laboratory code | Lab. date BP | Cal BP 2σ lower range | Cal BP 2σ upper range | Samples dated pre-1977 |
|---|---|---|---|---|---|
| Cueva de la Lechuza | LE-4276 | Charcoal | 2724 | 1890 | |
| Cueva de la Lechuza | LE-4270 | Charcoal | 3718 | 2850 | |
| Cueva de la Lechuza | LE-4273 | Charcoal | 2749 | 2181 | |
| Cueva de la Lechuza | LE-4279 | Charcoal | 2796 | 1996 | |
| Cueva de la Pintura | GD-1046 | Charcoal | 3158 | 2789 | 1973 |
| Cueva de la Pintura | GD-613 | Charcoal | 3242 | 2845 | 1973 |
| Cueva de la Pintura | GD-591 | Charcoal | 3341 | 2858 | 1973 |
| Cueva de la Pintura | GD-1039 | Charcoal | 2332 | 1996 | 1973 |
| Cueva de la Pintura | GD-614 | Charcoal | 2959 | 2742 | 1973 |
| Cueva de la Pintura | GD-601 | Charcoal | 3075 | 2770 | 1973 |
| Cueva del Perico I | GD-616 | Charcoal | 1376 | 1146 | 1972 |
| Cueva del Perico I | GD-1051 | Charcoal | 2146 | 1734 | 1972 |
| Cueva del Perico I | GD-617 | Charcoal | 1526 | 1294 | 1972 |
| Cueva Funche | SI-426 | Charcoal | 2352 | 1702 | 1966 |
| Cueva Funche | SI-429 | Charcoal | 4854 | 3994 | 1966 |
| Cueva Funche | SI-428 | Charcoal | 3828 | 2785 | 1966 |
| Cueva Funche | SI-427 | Charcoal | 3066 | 2112 | 1966 |
| Damayajabo | Y-1764 | Charcoal | 3697 | 3262 | |
| Damayajabo | Y-1994 | Charcoal | 1332 | 697 | |
| El Boniato (El Palmar) | Beta-148958 | Charcoal | 728 | 536 | |
| El Convento | GD-1053 | Charcoal | 686 | 546 | 1974 |
| El Convento | Unknown-7 | Charcoal | 507 | 338 | |
| El Guafe I | FS AC 2419 | Charcoal | 693 | 556 | |
| El Guafe I | FS AC 2420 | Charcoal | 534 | 476 | |
| El Morrillo | SI-353 | Charcoal | 686 | 498 | 1966 |
| El Paraiso | Unknown-8 | Charcoal | 1312 | 732 | |
| El Porvenir | Beta-148960 | Charcoal | 630 | 495 | |
| El Purial | UBAR-169 | Charcoal | 3644 | 2780 | |
| Esterito | SI-350 | Charcoal | 667 | 310 | 1965 |
| Esterito | SI-349 | Charcoal | 739 | 299 | 1965 |
| Herradura 1 | Beta-140075 | Marine shell | 1808 | 1438 | |
| Jorajuria | LE-1783 | Charcoal | 4827 | 4442 | |
| Jorajuria | LE-1784 | Charcoal | 4419 | 4152 | |
| Jorajuria | LE-1782 | Charcoal | 4241 | 3984 | |
| Jucaro | Beta-148949 | Charcoal | 728 | 548 | |
| La Escondida de Bucuey | Unknown-9 | Charcoal | 1292 | 682 | |
| La Guira | Beta-140077 | Terrestrial shell | 1407 | 1178 | |
| La Guira de Barajagua | SI-351 | Charcoal | 692 | 484 | 1965 |
| La Luz | Beta-93863 | Charcoal | 1342 | 1178 | |
| Laguna de Limones | SI-348 | Charcoal | 786 | 495 | 1964 |
| Levisa 1 (Far. de Lev.) | MC-860 | Charcoal | 5318 | 4828 | |
| Levisa 1 (Far. de Lev.) | GD-250 | Charcoal | 6288 | 5584 | 1973 |
| Levisa 1 (Far. de Lev.) | MC-859 | Charcoal | 5041 | 4520 | |
| Levisa 1 (Far. de Lev.) | GD-204 | Charcoal | 4150 | 3367 | |

| Site name | Laboratory code | Lab. date BP | Cal BP 2σ lower range | Cal BP 2σ upper range | Samples dated pre-1977 |
|---|---|---|---|---|---|
| Levisa 8 (Cueva S. Rita) | LE-2720 | Charcoal | 2858 | 2744 | |
| Levisa 8 (Cueva S. Rita) | LE-2718 | Charcoal | 2778 | 2623 | |
| Levisa 8 (Cueva S. Rita) | LE-2719 | Charcoal | 2313 | 2007 | |
| Levisa 8 (Cueva S. Rita) | LE-2717 | Charcoal | 2059 | 1876 | |
| Loma de la Campana | GD-1057 | Charcoal | 622 | 494 | 1972 |
| Loma de la Campana | GD-624 | Charcoal | 624 | 502 | 1972 |
| Loma de la Campana | GD-1056 | Charcoal | 670 | 518 | 1972 |
| Loma de la Forestal | SI-352 | Charcoal | 1066 | 686 | 1965 |
| Loma de Ochile | FS AC 2414 | Charcoal | 736 | 666 | |
| Loma de Ochile | FS AC 2415 | Charcoal | 693 | 556 | |
| Loma de Ochile | FS AC 2416 | Charcoal | 674 | 556 | |
| Loma de Ochile | FS AC 2417 | Charcoal | 663 | 544 | |
| Loma de Ochile | FS AC 2418 | Charcoal | 917 | 694 | |
| Los Buchillones | TO-8070 | Wood | 496 | 1 | |
| Los Buchillones | TO-7627 | Wood | 546 | 340 | |
| Los Buchillones | TO-7628 | Wood | 656 | 510 | |
| Los Buchillones | TO-8067 | Wood | 462 | 1 | |
| Los Buchillones | TO-7624 | Wood | 1334 | 1091 | |
| Los Buchillones | TO-7623 | Wood | 520 | 308 | |
| Los Buchillones | TO-7622 | Wood | 496 | 294 | |
| Los Buchillones | TO-7621 | Wood | 1404 | 1188 | |
| Los Buchillones | TO-7620 | Wood | 536 | 320 | |
| Los Buchillones | TO-7619 | Wood | 496 | 154 | |
| Los Buchillones | TO-7618 | Wood | 635 | 498 | |
| Los Buchillones | TO-8069 | Wood | 471 | 1 | |
| Los Buchillones | TO-8071 | Wood | 472 | 1 | |
| Los Buchillones | TO-8072 | Wood | 542 | 316 | |
| Los Buchillones | TO-7626 | Wood | 650 | 504 | |
| Los Buchillones | TO-8068 | Wood | 631 | 349 | |
| Los Buchillones | TO-7617 | Wood | 504 | 288 | |
| Los Buchillones | TO-7625 | Wood | 506 | 294 | |
| Los Chivos | Beta-140074 | Terrestrial shell | 1242 | 933 | |
| Los Chivos | Beta-140076 | Terrestrial shell | 2988 | 2722 | |
| Los Pedregales | GD-619 | Charcoal | 1286 | 927 | 1976 |
| Marien 2 | LV-2063 | Charcoal | 2293 | 1819 | |
| Marien 2 | LV-2062 | Charcoal | 924 | 553 | |
| Mejias | SI-347 | Charcoal | 1172 | 730 | 1965 |
| Mogote de la Cueva | Unknown-3 | Charcoal? | 961 | 742 | |
| Mogote de la Cueva | SI-424 | Charcoal | 1874 | 1278 | 1966 |
| Mogote de la Cueva | SI-425 | Charcoal | 957 | 299 | 1966 |
| Playita (Villa Clara) | Unknown-2 | Charcoal | 1282 | 1174 | |
| Potrero del Mango | Y-206 | Wood | 920 | 652 | |
| Potrero del Mango | Beta-148961 | Charcoal | 936 | 670 | |
| Potrero del Mango | Beta-148962 | Charcoal | 676 | 522 | |

*continued*

Table 4.2—*Continued*

| Site name | Laboratory code | Lab. date BP | Cal BP 2σ lower range | Cal BP 2σ upper range | Samples dated pre-1977 |
|---|---|---|---|---|---|
| Punta de Peque | Beta-93860 | Terrestrial shell | 1402 | 1187 | |
| San Benito | Beta-93851 | Terrestrial shell | 2140 | 1830 | |
| Vega del Palmar | Y-465 | Charcoal | 970 | 734 | Pre-1959 |
| Ventas de Casanova | FS AC 2421 | Charcoal | 503 | 310 | |
| Ventas de Casanova | FS AC 2424 | Charcoal | 542 | 496 | |
| Ventas de Casanova | FS AC 2422 | Charcoal | 530 | 321 | |
| Ventas de Casanova | FS AC 2423 | Charcoal | 498 | 288 | |
| Victoria I | LC-H 565 | Charcoal | 961 | 742 | |
| Victoria I | LC-H 1034 | Charcoal | 2338 | 1816 | |
| Victoria I | LC-H 1035 | Charcoal | 1518 | 1272 | |

brated dates were put into millennia BP to provide a broad chronological sequence for radiocarbon chronologies (figure 4.6).

The small sample size of calibrated radiocarbon dates, in addition to the potential methodological issues of date comparison, limits the degree to which meaningful interpretations can be made based on observations in spatial patterns of the dates. However, initial observations at least provide some tentative hypotheses that can then be challenged and further investigated through targeted future research. One observation that can be made in the spatial patterns of calibrated radiocarbon dates is that the seven earliest radiocarbon dates come from four sites that are all located in relation to the north coast of Cuba. These radiocarbon dates range between 6300 and 4450 cal BP. The first radiocarbon dates to appear close to the south coast are from the sites of Cueva Funche (4850–3990 cal BP) and Damajayabo (3700–3260 cal BP) in the west and east of the island, respectively. The earliest radiocarbon determinations from an offshore island come from Cave 4 (1290–740 cal BP), and Cave 1 (970–680 cal BP) from Punta del Este, Isla de la Juventud. The sites with radiocarbon determinations from the last one thousand years appear to be distributed throughout eastern and central Cuba. There are no radiocarbon determinations for this period from the westernmost area of Cuba. This distribution appears similar to the distribution of sites with indigenous ceramics and sites classified as agroalfarero. Further radiocarbon determinations are required to identify whether these patterns reflect a small sample of radiocarbon determinations or if they are archaeologically significant.

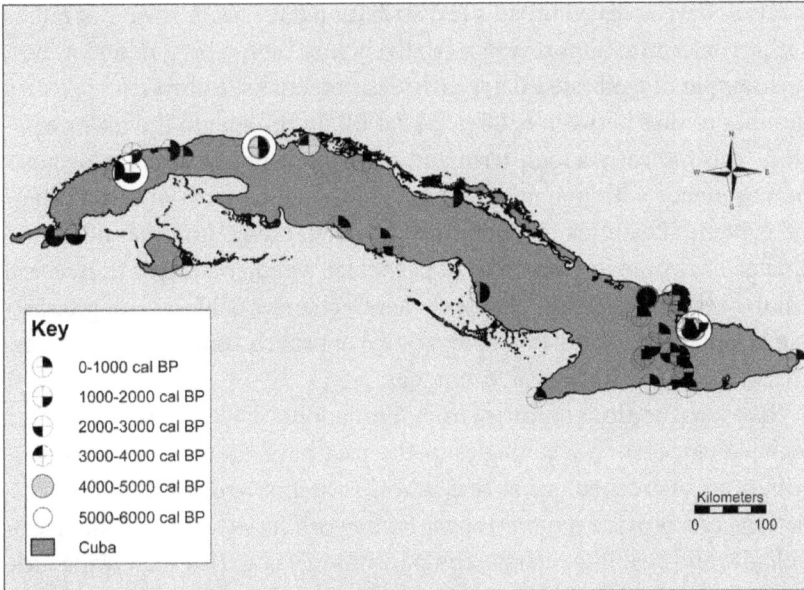

Figure 4.6. Spatial distribution of existing radiocarbon dates in Cuba, based on the upper range of calibrated radiocarbon determinations.

Collating all of the available information on radiocarbon determinations in Cuba, and providing a platform for projecting their spatial locations through time, can facilitate discussion of their relevance in understanding the prehistoric settlement of Cuba. One hundred and forty radiocarbon dates for a country as large as Cuba, and with such a long period of human occupation, is a comparatively small sample. Methodological issues concerning the comparison between radiocarbon dates discussed here also need to be taken into account before interpreting the archaeological significance of comparisons between calibrated radiocarbon dates.

## Conclusions

This study provides a summary of archaeological information collected from 1,061 archaeological sites in Cuba. The creation of a database of archaeological sites, spatially projected in GIS, provides a useful framework for studies of pre-Columbian archaeology in Cuba. This study has illustrated how national studies of site distribution in Cuba require the collection of large data sets. Spatial studies of archaeological material and site location have revealed broad site distribution patterns. A review of radio-carbon determinations reveals a relatively small corpus of dates. However, this sample of calibrated dates provides potential site chronologies for a number of sites between 6300–5750 cal BP up through to the historic period. This indicates a long-term and continuous pre-Columbian occupation sequence with potential evidence of spatial patterns in archaeological evidence changing through time. Understanding processes of culture change over time in Cuba, and in particular whether changes in material culture reflect changes in people or merely changes of lifestyle of preexisting populations, requires a comparative approach using a variety of archaeological and biological techniques.

Two key questions that arise from this national database are (1) whether archaeological survey strategies in the past have resulted in biased distributions of archaeological sites and (2) whether radiocarbon determinations can provide a more robust framework for establishing site chronologies and enabling intersite comparisons in Cuba. This islandwide scale of analysis allows the observation of macro-scale patterns that can then be investigated through informed archaeological research at a local scale where targeted research questions can be addressed in greater detail. Current research being carried out by the Cuban Ministry of Science, Technology and Environment and by University College London is attempting to address some of these issues through collaborative archaeological field-work in a local case-study area in north-central Cuba (Cooper et al. 2006; Valcárcel Rojas et al. 2006).

In this chapter, I have attempted to collate archaeological data in Cuba and study spatial patterns in site distribution using GIS applications. This study has recorded the names and locations of pre-Columbian archaeological sites in Cuba, details of artifact assemblages recovered from each site, current site classifications based on existing archaeological frameworks, and radiocarbon determinations from archaeological sites that could be used to construct site chronologies.

Current research is directed not only at expanding this newly established database of Cuban archaeology and using different GIS applications to model spatial patterns but also at answering research questions that arise from these analyses. Research initiatives that have arisen from this study of site distribution patterns and site chronologies and that form part of ongoing collaborative archaeological research in Cuba include (1) examining spatial patterns in site distribution through systematic archaeological surveys of targeted case-study areas; (2) reviewing existing site classifi-cations of known sites in Cuba through studies of artifact assemblages recovered during new archaeological excavations; and (3) establishing the relative chronologies of a group of sites, using radiocarbon dating, in order to generate a more robust temporal framework for studying site interac-tion and culture change through time.

In sum, the development of a nationwide archaeological database that includes locational and chronological data is a critical step toward docu-menting the rich cultural heritage of Cuba and developing more-sophis-ticated research questions about pre-Columbian settlement patterns. It is hoped that as ongoing archaeological research in Cuba begins to address these research questions, the origins and cultural development of Cuba's early populations will be better illuminated.

# 5

# A Morphometric Approach to Taíno
# Biological Distance in the Caribbean

ANN H. ROSS AND DOUGLAS H. UBELAKER

## Abstract

Dispersal hypotheses in the Caribbean are evaluated using both traditional craniometric and modern geometric morphometric methods. A study of craniofacial shape variation was conducted among precontact Taíno (Arawak) groups from Cuba, Hispaniola, Puerto Rico, and Jamaica and precontact groups from Florida, Mexico, Panama, Colombia, Venezuela, and Ecuador. The between-group variation and the degree of among-group differentiation were tested using Mahalanobis $D^2$. Results demonstrate morphological similarity among samples from Puerto Rico, the Dominican Republic, and Jamaica but document the distinctiveness of the Cuban sample. The MANOVA (Wilks' $\Lambda$ = .005; F value = 4.13; $df$ = 160, 641.62; $p < .0001$) procedure detected significant group differences. Interestingly, Cuba is significantly different than the rest of the Taíno series, suggesting a dissimilar origin and at least two separate migration routes. The results of this study suggest that New World population diffusion may not clearly fit into traditional spatial migratory models. The research reported here is part of a larger effort to construct a database documenting patterns of human cranial variation within Latin America prior to European contact.

## Resumen

Las hipótesis de la dispersión en el Caribe son evaluadas utilizando la metodología tradicional de craniometría y modernos de la geometría morfométrica. Este studio de la variación craniofacial fue realizado entre grupos Taínos pre-Colombinos de Cuba, de Hispaniola, de Puerto Rico, y de Jamaica y grupos pre-Colombinos de la Florida, México, Panamá, Colombia, Venezuela, y del Ecuador. La variación morfométrica y el grado de diferenciación biologica fueron probados usando un análisis multivariado de los valores $D^2$ de Mahalanobis como distancias morfométricas entre pares de muestras. Los resultados demuestran la similitud morfológica entre muestras de Puerto Rico, la República Dominicana, y Jamaica, pero documentan la peculiaridad de la muestra cubana. El procedimiento de MANOVA (Wilks' $\Lambda$ = .005; F value = 4.13; $df$ = 160, 641.62; $p$ < .0001) discernió una diferencia significativa entre los grupos. Curiosamente, Cuba es apreciablemente diferente del resto de la serie de Taínos, sugiriendo un origen diferente y más de una ruta migratoria. Los resultados de este estudio sugieren que los modelos migratorios espaciales tradicionales no son adecuados para explicar el poblamiento de las Américas. Esta investigación forma parte de un esfuerzo más grande de construir una base de datos que documenta las pautas de la variación craneal humana dentro de Iberoamérica antes del contacto europeo.

*   *   *

The timing and nature of the peopling of the New World have been of great historical interest. Although the bulk of the evidence suggests general origins from northeastern Asia via Beringia, specifics have remained elusive, generating considerable discussion, controversy, and variation in interpretation. Much of this discussion centers on the antiquity and number of migrations, as well as the routes traveled by the early Americans. Evidence bearing on these questions stems from multiple sources but especially archaeology, dating of well-documented specimens, linguistic analysis, and biological anthropology. Within biological anthropology, interpretations center on analysis of diversity within recent populations, including molecular studies and research in human skeletal biology, which have recently incorporated perspectives from ancient DNA.

Craniometric studies have long been employed within anthropology to document aspects of human variation and to assess population relationships (Ubelaker and Jantz 1986). The availability of high-speed computers and advances in statistical approaches have greatly enhanced the potential use of such studies, especially to assemble large databases of measurements and to use them in sophisticated ways to examine sample patterns and relationships (Howells 1973; Ross et al. 2002b).

Despite the importance of data from Latin America in holistic interpretation of the issues discussed above, craniometric studies from the region have lagged behind those of North America. Although important regional studies have been conducted (for example, Cocilovo and Rothhammer 1990, 1996, 1999; Rothhammer et al. 1982, 1984; Varela and Cocilovo 1999, 2000), many researchers have utilized the studies of W. W. Howells (1973) for perspective on craniometric variation, which includes only one Peruvian sample to represent Latin America. The research reported here stems from a broader effort (Ross 2004; Ross et al. 2002a) to build a large craniometric database from precontact samples throughout Latin America. Hopefully, such a database will provide a more realistic view of the extent of variation in the region, a context in which to judge morphology of the few early specimens that have been recovered and insight into the dynamics of past migration events.

The most recognized Antillean Saladoid dispersal hypothesis is a direct jump by agriculturalists from South America, followed by dispersal into the Lesser Antilles and westward (Keegan 1995; Moreira de Lima 1999). The origins of the Taíno, or Arawakan speakers, have been traced to the Orinoco River in Venezuela (Rouse 1992a; Keegan 1992; Callaghan 2001). Due to the shortage of relatively well-preserved skeletal material, and possibly to researcher bias for investigating Paleoamerican settlement patterns rather than more recent migration events, the Caribbean has been overlooked by most biological anthropologists. Therefore, this investigation attempts to evaluate possible routes of population dispersals in the Caribbean by examining biological shape variation in precontact Caribbean Taíno groups and comparing these to precontact North and South American groups using both traditional craniometric and modern geometric morphometric methods.

## Traditional Craniometrics

### Sample

The database reported here consists of 109 individuals from eight precontact samples originating from the Caribbean and elsewhere in Latin America (table 5.1). The traditional craniometric analysis differs from the three-dimensional analysis in that it includes a precontact sample from

Table 5.1. Sample Composition for the Traditional Craniometric Analysis

| Sample name | N | Provenience |
|---|---|---|
| Colombia | 5 | Precontact, American Museum of Natural History, New York City |
| Cuba | 17 | (ca. AD 800–1500) Museo de Montane, Havana, Cuba |
| Dominican Republic | 17 | (ca. AD 800–1500) National Museum of Natural History, Smithsonian Institution, Washington, D.C. |
| Ecuador | 20 | (ca. AD 730–1500) National Museum of Natural History, Smithsonian Institution, Washington, D.C. |
| Jamaica | 7 | (ca. AD 800–1500) National Museum of Natural History, Smithsonian Institution, Washington, D.C. |
| Mexico | 30 | Carl Lumholtz Collection, American Museum of Natural History, New York City |
| Puerto Rico | 9 | (ca. AD 800–1542) American Museum of Natural History, New York City |
| Venezuela | 4 | Precontact, American Museum of Natural History, New York City |

Ecuador ($n$ = 20) not used in the geometric morphometric analysis. Only adult crania were included, and males and females were pooled in order to include all of the observed biological variation within the population and to enhance the sample sizes. Because preservation is an issue in tropical environments, some group samples are small, for preservation of facial landmarks is required for morphometric analyses though it may not be necessary for other analyses, such as DNA. The sample from Tarasco is from the Carl Lumholtz Collection housed at the American Museum of Natural History. All that is known about this sample is that Carl Lumholtz's last expedition to the state of Michoacan, Mexico, to collect crania from a prehistoric Tarasco site called "El Palacio" took place between 1894 and 1897 (Lumholtz and Hrdlička 1898). The sample from Ecuador is from coastal urn burials from the Ayalán site (Ubelaker 1981). The Cuban sample housed at the Museo de Montane in Havana is from various sites, which are known to be Taíno because of the associated ceramics and their widely known practice of fronto-occipital deformation. The only information regarding the samples from Hispaniola, Jamaica, and Puerto Rico is that they are Taíno or Arawakan speakers. Specific information regarding the provenience of the samples from Colombia and Venezuela

Table 5.2. Measurement Description

| Measurement | Description |
| --- | --- |
| Minimum frontal breadth (WFB) | Martin 1956: 457, #9; Moore-Jansen et al. 1994: 53, #11 |
| Nasal height (NLH) | Howells 1966: 6, #14; Martin 1956: 479, #55; Moore-Jansen et al. 1994: 54, #13 |
| Nasal breadth (NLB) | Howells 1973: 176; Martin 1956: 479, #54; Moore-Jansen et al. 1994: 54, #14 |
| Interorbital breadth (using dacryon) (DKB) | Martin 1956: 477, #49a; Moore-Jansen et al. 1994: 55, #18 |
| Face breadth (ectoconchion) (EKB) | Howells 1973: 178; Moore-Jansen et al. 1994: 55, #17 |
| Orbit height (OBH) | Martin 1956: 478, #52; Moore-Jansen et al. 1994: 55, #16 |
| Orbit breadth (OBB) | Howells 1973: 175; Martin 1956: 477–78, #51a; Moore-Jansen et al. 1994: 55, #15 |
| Foramen magnum length (FOL) | Martin 1956: 455, #7; Moore-Jansen et al. 1994: 56, #22 |
| Foramen magnum breadth (FOB) | Martin 1956: 459, #16; Moore-Jansen et al. 1994: 57, #23 |
| Upper facial breadth (UFBR) | Martin 1956: 475, #43; Moore-Jansen et al. 1994: 54, #12 |

is unknown, however. Sex-related variation is assumed to be negligible within each population in among-population comparisons (Sardi Marina et al. 2005). The ten measurements utilized are summarized in table 5.2 and were available for all of the crania included in the study.

Statistics

To account for size effects, size and shape variables were calculated from raw measurements utilizing methodology proposed by Mosimann and James (1979) and Darroch and Mosimann (1985). In this approach, *size* is defined as the geometric mean (GM) of all cranial shape variables. Each raw variable is divided by the GM to create shape variables, simple ratios of the geometric mean that no longer directly express size (Falsetti et al. 1993) but may be linked to size. Thus, while this approach does not entirely dispense with size issues, it offers improved insight through the analysis of "geometric similarity" among the samples examined.

A one-way analysis of variance (ANOVA) was conducted on the size variable to test the null hypothesis that mean size is not significantly different among the samples examined. Canonical variates (linear combinations of predictor variables that summarize between-population variation) were then calculated from the altered shape variables to establish distances among the samples. A Pearson correlation analysis was also conducted to measure the strength of the relation between the GM and the canonical axes.

To establish the degree of differentiation among the samples, Mahalanobis $D^2$ or generalized squared distance was employed; this represents a function of the group means and pooled variances and covariances (Afifi and Clark 1996). This statistic is used to test whether sample centroids are significantly different. The multivariate analyses were performed using the SAS system for Windows Version 9.1.3.

## Geometric Morphometrics

### Sample

The sample used in this study totaled 108 individuals from the Caribbean and Latin America. The Caribbean samples have all been identified archaeologically as Taíno. The name "Taíno" refers to indigenous peoples of the Greater Antilles in the Caribbean with a developed agricultural subsistence economy (Martinón-Torres et al. 2007). In Cuba, Taíno sites are almost exclusively identified through ceramic typologies (Martinón-Torres et al. 2007). The sample composition is presented in table 5.3. This part of the analysis also included a precontact sample from Panama. For this project, we selected eighteen homologous facial landmarks that should reflect the among-group variation (table 5.4 and figure 5.1). Only facial landmarks were utilized, because the Taínos practiced intentional cranial modification. Because of the increased range of morphological variation produced by intentional cranial reshaping, many skeletal series have not been examined in this manner. However, Cocilovo (1973, 1975, and 1978) in three separate studies of modified Andean and lowland crania found that facial measurements were not significantly distorted. Antón (1989) did find significant differences between modified and unmodified crania. However, the research design was flawed in that the deformed and undeformed samples used came from different populations and not the same

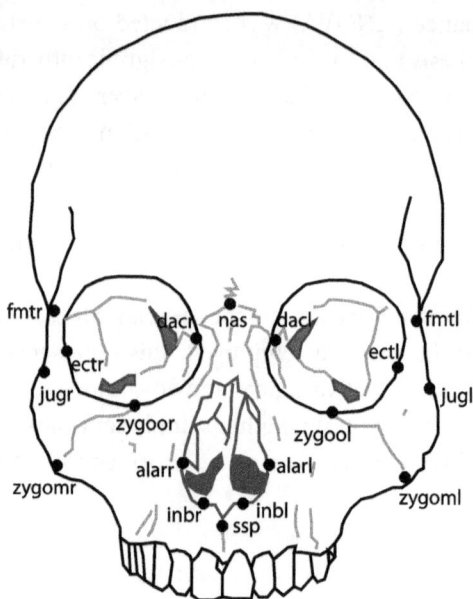

Figure 5.1. Location of facial landmarks.

Table 5.3. Sample Composition for the Geometric Morphometric Analysis

| Group | N | Provenience |
|---|---|---|
| Colombia | 5 | Precontact, American Museum of Natural History |
| Cuba | 17 | (ca. AD 800–1500) Museo de Montane, Havana, Cuba |
| Hispaniola | 15 | (ca. AD 800–1500) National Museum of Natural History |
| Jamaica | 7 | (ca. AD 800–1500) National Museum of Natural History |
| Mexico | 31 | Tarasco, American Museum of Natural History |
| Panama | 6 | Precontact, Patronato Panama Viejo |
| Puerto Rico | 9 | (ca. AD 800–1542) American Museum of Natural History |
| Florida | 14 | (ca. AD 1300–1400) National Museum of Natural History |
| Venezuela | 4 | Precontact, American Museum of Natural History |

Table 5.4. Facial Landmarks Used in the Geometric Morphometric Analysis

| | |
|---|---|
| Alare right/left (alarr/alarl) | Nasal inferior border right/left (inbr/inbl) |
| Dacryon right/left (dacr/dacl) | Nasion (nas) |
| Ectochonchion right/left (ectr/ectl) | Subspinale (ssp) |
| Frontomalare temporale right/left (fmtr/fmtl) | Zygomaxillare right/left (zygomr/zygoml) |
| Jugale right/left (jugr/jugl) | Zygoorbitale right/left (zygoor/zygool) |

population. Therefore, the results most likely depict population variation rather than significant structural changes due to cranial modification. The appropriateness of using facial landmark data for population distance studies is forthcoming and will be addressed in another manuscript. A Microscribe G2X digitizer was utilized to obtain the $x$, $y$, and $z$ coordinates for each landmark using the program ThreeSkull, written by Steve Ousley.

## Analysis

Because shape analysis requires that data be invariant to location, orientation, and scale, raw coordinate data cannot be directly compared between specimens, as each data set was collected in its own coordinate system. Thus, the data must be translated and rotated to a common coordinate system and scaled to a common size before they can be directly comparable and subjected to multivariate tests. To undertake these transformations, a generalized procrustes analysis (or GPA) was used to minimize the sum of squared distances between landmarks of each skull and those of an iteratively computed mean. The GPA superimposition was performed using the program Morpheus et al., written by Dennis Slice (1998). The resultant shape variables were then utilized in the following multivariate analyses. A principal component analysis (PCA) of the covariance matrix was conducted on the GPA-transformed variables to reduce the dimensionality of the data or as a variable reduction procedure to meet the requirements of the parametric test. Minimally, one more specimen is needed than the number of dimensions. Next, a multivariate analysis of variance (MANOVA) was performed using the dimension-reducing PCA scores to test for among-group differences. The degree of differentiation among the groups was evaluated using Mahalanobis $D^2$, which is a function of the group means and the pooled variances and covariances of the principal components scores (Afifi and Clark 1996). Additionally, an UPGMA (unweighted pair group method with arithmetic mean) clustering analysis was conducted on the Mahalanobis $D^2$ distance matrix based on the first twenty principal components, and a separate clustering procedure was performed based on the first fourteen principal component scores (Sneath and Sokal 1973). The multivariate analyses were performed using the SAS system for Windows Version 9.1.3.

## Results

Traditional Craniometrics

Table 5.5 presents the summary statistics for each measurement of each sample. The generalized squared Mahalanobis distance results ($D^2$) reveal the extent of differentiation within groups and are summarized in table 5.6. Although these results present considerable variation in the level of statistical significance, they generally indicate heterogeneity in many of the between-sample comparisons. The three significant canonical axes are presented in table 5.7. These reveal that approximately 46 percent of the among-group shape variation is accounted for by CAN1, 27 percent by CAN2, and 18 percent by CAN3. Total canonical structure for CAN1, CAN2, and CAN3 representing the correlation between the original variables and the canonical variates is presented in table 5.8. Both the Cuba and the Ecuador samples demonstrate marked distance from the others. The Puerto Rico sample shows closest affinity to the Dominican Republic, followed by Mexico, Jamaica, and Ecuador. Despite the close proximity, Cuba shows the greatest distance. In contrast, Cuba shows the closest distance with the Ecuador sample, despite the fact that the geographic distance is greatest between the sites of origin of these samples. The distances between the Cuban sample and all others are generally consistent with geography, except for greatest biological distance with the Jamaica sample. The Dominican Republic sample logically links closest with Puerto Rico, followed by Jamaica, but then the geographically distant Ecuador sample is more similar than Mexico, Venezuela, Cuba, or Colombia. Jamaica links closely with its Caribbean neighbors the Dominican Republic and Puerto Rico. These close linkages are followed by Mexico, Venezuela, Cuba, and Colombia. The distance with Cuba again underscores the distinctiveness of the Cuban sample. The Mexican sample shows close linkage with Colombia, followed by Puerto Rico and the Dominican Republic. The Cuban sample shows the greatest distance. The Venezuelan sample links most closely with its South American neighbor Colombia, followed by Mexico and Puerto Rico. The sample from Colombia shows the shortest biological distance with Mexico, followed by Venezuela and Puerto Rico. The greatest biological distance is with Cuba, Ecuador, and Jamaica, in that order. Finally, the Ecuador sample shows the strongest morphological linkage (shortest distance) with three Caribbean samples (Puerto Rico, the Dominican Republic, and Cuba). The greatest biological distance is displayed

Table 5.5. Traditional Craniometric Summary Statistics for Each Sample

| Sample name | Minimum frontal breadth (wfb) | | | | Nasal height (nlh) | | | | Nasal breadth (nlb) | | | |
|---|---|---|---|---|---|---|---|---|---|---|---|---|
| | Mean | S.D. | Min. | Max. | Mean | S.D. | Min. | Max. | Mean | S.D. | Min. | Max. |
| Colombia | 88.40 | 3.71 | 83.00 | 93.00 | 51.00 | 2.83 | 48.00 | 54.00 | 26.00 | 2.00 | 24.00 | 29.00 |
| Cuba | 94.06 | 4.57 | 83.00 | 101.00 | 51.71 | 3.41 | 46.00 | 58.00 | 25.47 | 1.50 | 23.00 | 28.00 |
| Dominican Republic | 95.00 | 4.86 | 87.00 | 102.00 | 49.06 | 2.63 | 45.00 | 55.00 | 25.82 | 1.91 | 23.00 | 30.00 |
| Ecuador | 92.20 | 3.81 | 85.00 | 99.00 | 47.55 | 3.10 | 41.00 | 52.00 | 23.65 | 1.60 | 21.00 | 27.00 |
| Jamaica | 99.14 | 4.02 | 93.00 | 104.00 | 52.71 | 2.87 | 50.00 | 57.00 | 27.14 | 1.77 | 24.00 | 29.00 |
| Mexico | 92.10 | 3.34 | 86.00 | 99.00 | 51.17 | 3.68 | 43.00 | 58.00 | 26.77 | 1.91 | 23.00 | 30.00 |
| Puerto Rico | 94.44 | 4.10 | 88.00 | 100.00 | 50.67 | 3.24 | 46.00 | 55.00 | 26.78 | 1.99 | 24.00 | 30.00 |
| Venezuela | 93.50 | 2.65 | 91.00 | 97.00 | 53.50 | 7.85 | 42.00 | 59.00 | 25.75 | 3.86 | 22.00 | 30.00 |

| Sample name | Orbit height (obh) | | | | Orbit breadth (obb) | | | | Interorbital breadth (dkb) | | | |
|---|---|---|---|---|---|---|---|---|---|---|---|---|
| | Mean | S.D. | Min. | Max. | Mean | S.D. | Min. | Max. | Mean | S.D. | Min. | Max. |
| Colombia | 34.40 | 2.19 | 31.00 | 37.00 | 37.80 | 1.79 | 35.00 | 40.00 | 22.00 | 1.58 | 20.00 | 24.00 |
| Cuba | 39.88 | 2.20 | 35.00 | 44.00 | 41.12 | 1.83 | 38.00 | 45.00 | 21.59 | 2.55 | 18.00 | 26.00 |
| Dominican Republic | 36.00 | 1.90 | 34.00 | 41.00 | 37.94 | 1.48 | 36.00 | 41.00 | 22.35 | 2.23 | 19.00 | 27.00 |
| Ecuador | 35.50 | 1.61 | 33.00 | 39.00 | 38.40 | 1.50 | 35.00 | 41.00 | 19.90 | 1.86 | 17.00 | 23.00 |
| Jamaica | 36.57 | 1.51 | 35.00 | 39.00 | 38.43 | 1.27 | 36.00 | 40.00 | 22.43 | 2.64 | 17.00 | 24.00 |
| Mexico | 35.10 | 2.16 | 30.00 | 39.00 | 39.03 | 1.59 | 36.00 | 42.00 | 21.47 | 1.53 | 19.00 | 24.00 |
| Puerto Rico | 35.67 | 2.50 | 33.00 | 40.00 | 39.00 | 1.66 | 37.00 | 41.00 | 21.33 | 2.87 | 17.00 | 27.00 |
| Venezuela | 36.00 | 2.94 | 32.00 | 39.00 | 39.00 | 1.63 | 37.00 | 41.00 | 22.75 | 1.50 | 21.00 | 24.00 |

continued

Table 5.5—Continued

| Sample name | Face breadth (ekb) | | | | Foramen magnum length (fol) | | | | Foramen magnum breadth (fob) | | | |
|---|---|---|---|---|---|---|---|---|---|---|---|---|
| | Mean | S.D. | Min. | Max. | Mean | S.D. | Min. | Max. | Mean | S.D. | Min. | Max. |
| Colombia | 94.40 | 2.88 | 90.00 | 97.00 | 35.60 | 1.52 | 34.00 | 37.00 | 29.60 | 2.07 | 27.00 | 32.00 |
| Cuba | 98.94 | 2.68 | 95.00 | 103.00 | 36.06 | 2.28 | 32.00 | 41.00 | 32.24 | 2.02 | 29.00 | 35.00 |
| Dominican Republic | 96.12 | 3.06 | 91.00 | 102.00 | 36.41 | 3.02 | 31.00 | 44.00 | 31.65 | 2.47 | 28.00 | 37.00 |
| Ecuador | 93.75 | 3.51 | 87.00 | 99.00 | 33.85 | 3.20 | 25.00 | 39.00 | 30.65 | 4.45 | 20.00 | 38.00 |
| Jamaica | 96.86 | 3.44 | 91.00 | 101.00 | 35.71 | 0.95 | 34.00 | 37.00 | 31.86 | 1.35 | 30.00 | 34.00 |
| Mexico | 96.83 | 3.13 | 91.00 | 103.00 | 36.30 | 2.56 | 32.00 | 43.00 | 30.03 | 1.73 | 27.00 | 34.00 |
| Puerto Rico | 96.56 | 3.47 | 90.00 | 101.00 | 35.22 | 2.44 | 32.00 | 39.00 | 31.78 | 2.99 | 27.00 | 37.00 |
| Venezuela | 98.25 | 4.57 | 95.00 | 105.00 | 37.25 | 1.26 | 36.00 | 39.00 | 31.75 | 2.22 | 29.00 | 34.00 |

| Sample name | Upper facial breadth (ufbr) | | | |
|---|---|---|---|---|
| | Mean | S.D. | Min. | Max. |
| Colombia | 101.40 | 4.51 | 96.00 | 107.00 |
| Cuba | 103.29 | 3.74 | 94.00 | 108.00 |
| Dominican Republic | 103.29 | 4.63 | 95.00 | 112.00 |
| Ecuador | 101.20 | 3.69 | 95.00 | 106.00 |
| Jamaica | 105.57 | 4.54 | 97.00 | 111.00 |
| Mexico | 104.30 | 3.51 | 99.00 | 113.00 |
| Puerto Rico | 103.89 | 4.40 | 99.00 | 112.00 |
| Venezuela | 106.75 | 4.86 | 102.00 | 113.00 |

**Table 5.6. Generalized Squared Mahalanobis Distance Using Principal Component Scores ($D^2$), showing the Degree of Differentiation among the Samples**

| Group | Colombia | Cuba | Dominican Republic | Ecuador | Jamaica | Mexico | Puerto Rico | Venezuela |
|---|---|---|---|---|---|---|---|---|
| Colombia | 0 | | | | | | | |
| Cuba | 9.41110* | 0 | | | | | | |
| Dominican Republic | 6.91128* | 6.11295 | 0 | | | | | |
| Ecuador | 9.92356* | 5.94740 | 4.14478* | 0 | | | | |
| Jamaica | 12.26129* | 11.65083 | *3.10703* | 8.59869 | 0 | | | |
| Mexico | *1.48125* | 9.82039 | 4.74992 | 6.32731 | 9.15155 | 0 | | |
| Puerto Rico | 5.24542 | 7.25729* | *1.85392* | 3.57405* | *3.46159* | *2.51864* | 0 | |
| Venezuela | *1.49268* | 9.20914* | *5.76777* | 6.47003* | 11.03797* | *2.28757* | *4.89534* | 0 |

Note: All groups significant at $p < .0001$ except those marked *, which are significant at $p < .05$. Figures in italics indicate no significance.

**Table 5.7. Significant Canonical Axes for Transformed Variables**

| No. Eigenvalue | Difference | Proportion | Cumulative | Likelihood approximate Ratio | F value | Num. df | Den. df | Pr. > F |
|---|---|---|---|---|---|---|---|---|
| 1.2506 | 0.5223 | 0.4615 | 0.4615 | 0.13739104 | 3.15 | 70 | 543.26 | <.0001 |
| 0.7283 | 0.2291 | 0.2687 | 0.7302 | 0.30921044 | 2.30 | 54 | 478.8 | <.0001 |
| 0.4991 | 0.3791 | 0.1842 | 0.9144 | 0.53439743 | 1.59 | 40 | 412.53 | 0.0145 |

**Table 5.8. Total Canonical Structure for Transformed Variables**

| Variable | CAN1 | CAN2 | CAN3 |
|---|---|---|---|
| VFB | 0.307549 | -0.688414 | 0.419998 |
| ILH | -0.260543 | 0.165329 | -0.172430 |
| LB | -0.534555 | -0.174141 | -0.302484 |
| BH | 0.670624 | 0.329232 | -0.046561 |
| BB | 0.299777 | 0.547953 | 0.416924 |
| KB | -0.186242 | -0.147007 | -0.354096 |
| KB | 0.025355 | 0.391340 | 0.605323 |
| OL | -0.260756 | 0.066545 | -0.077562 |
| OB | 0.337289 | -0.132433 | 0.111028 |
| FBR | -0.339377 | -0.202564 | 0.745535 |

with Colombia, a sample originating in a country sharing a common border with the country of Ecuador.

Geometric Morphometrics

The MANOVA procedure detected significant group differences (Wilks' $\Lambda$ = .005; F value = 4.13; $df$ = 160, 641.62; $p$ < .0001). Difference between groups can be illustrated by plots showing difference vectors originating at the landmarks of one specimen and going to the position of the same landmarks on the other specimen. The difference vectors show the direction and magnitude of difference between one mean form and another. Figures 5.2 and 5.3 depict an anterior view of the mean configuration for Cuba and Florida and for Cuba and Panama, which were found to be dissimilar. The dissimilarity is suggested by the length of the vectors (which have been magnified by a power of four), but one must keep in mind that the statistical significance of differences between groups is a function of both mean differences and shape variability. Cuba is also significantly different from the rest of the Caribbean Taínos (figure 5.4). These differences can largely be observed in the lower nasal border and at zygomaxillare. However, similarity of the orbital region can be observed by the shortness of the vectors at zygoorbitale, dacryon, and nasion. The similarity between Colombia and Mexico is shown in figure 5.5, which is suggested by the

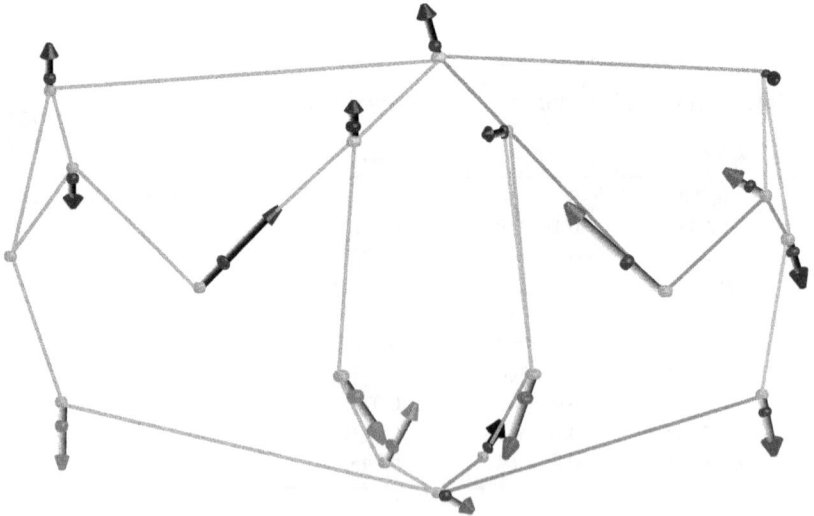

Figure 5.2. Mean cranial landmark differences, shown as vectors, from Cuba (light gray) to Florida (dark gray). Vectors magnified ×4.

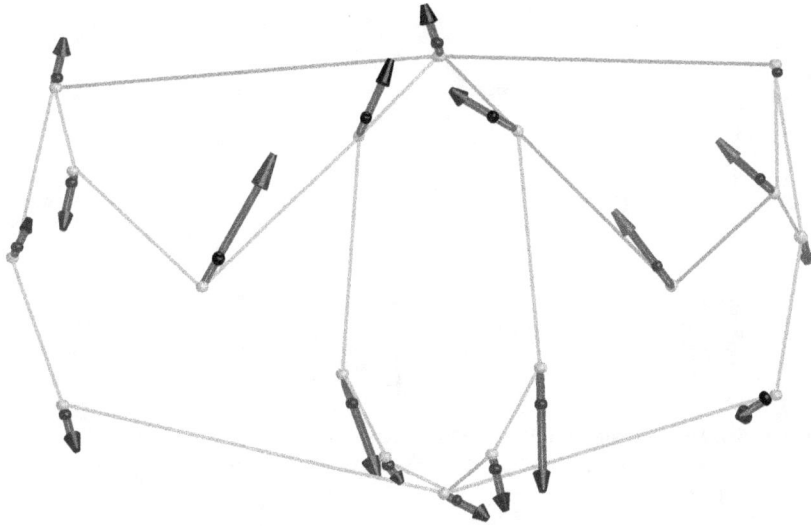

Figure 5.3. Mean cranial landmark differences, shown as vectors, from Cuba (light gray) to Panama (dark gray). Vectors magnified ×4.

shortness of the vectors. Interestingly, this is consistent with earlier findings of possible contact between Mexico and Ecuador (Ross et al. 2002a; Callaghan 2003b). The distinctiveness of Panamanian and Floridian crania, seen here in lateral view, is shown by the long vectors (figure 5.6). Table 5.9 presents the Mahalanobis squared distances, which confirm the similarities between Colombia and Mexico; Hispaniola and Jamaica; and Jamaica, Puerto Rico, and Venezuela. Interestingly, Cuba is considerably different from the rest of the Taíno and American groups. The p-values are presented on the upper right diagonal and the distances on the lower diagonal. These analyses were based on all twenty PCA scores.

The cluster analysis of the distances from all twenty PCA scores (accounting for approximately 88.5 percent of the total variation) produced some interesting results (figure 5.7). Three distinct groupings are observed. One cluster includes Mexico, Puerto Rico, and Colombia; another includes Hispaniola and Jamaica; and Florida and Panama cluster together. Venezuela is similar to the first two clusters, as is demonstrated by the short linkage distance. Notably, Cuba branches away from the other Caribbean Taíno and South American groups. We then performed a cluster analysis based on the first fourteen principal components (accounting for approximately 80 percent of the total variation), as they contained most of the variation present that captured the regional information (figure 5.8). The

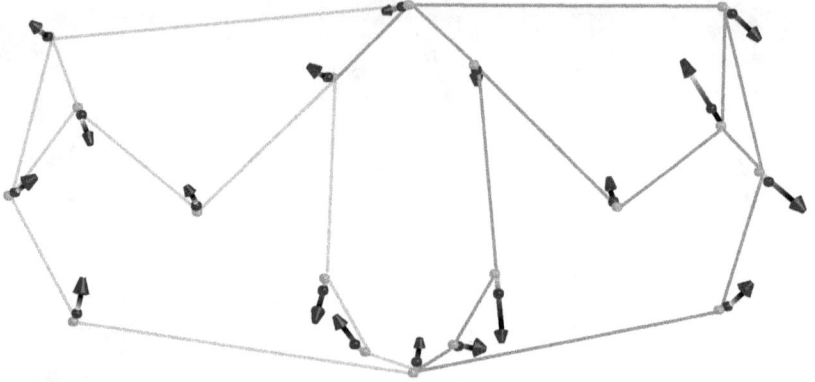

Figure 5.4. Mean cranial landmark differences, shown as vectors, from Cuba (light gray) to combined means for Hispaniola + Jamaica + Puerto Rico (dark gray). Vectors magnified ×4.

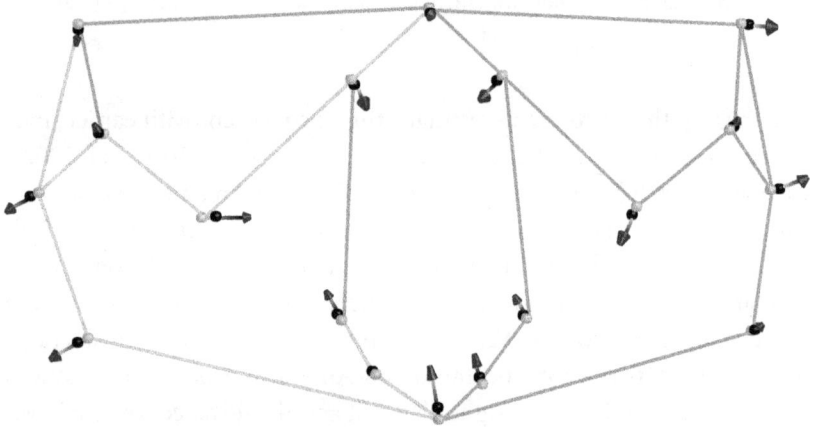

Figure 5.5. Mean cranial landmark differences, shown as vectors, from Colombia (light gray) to Mexico (dark gray). Vectors magnified ×4.

Figure 5.6. Mean cranial landmark differences, shown as vectors, from Panama (light gray) to Florida (dark gray). Vectors magnified ×4.

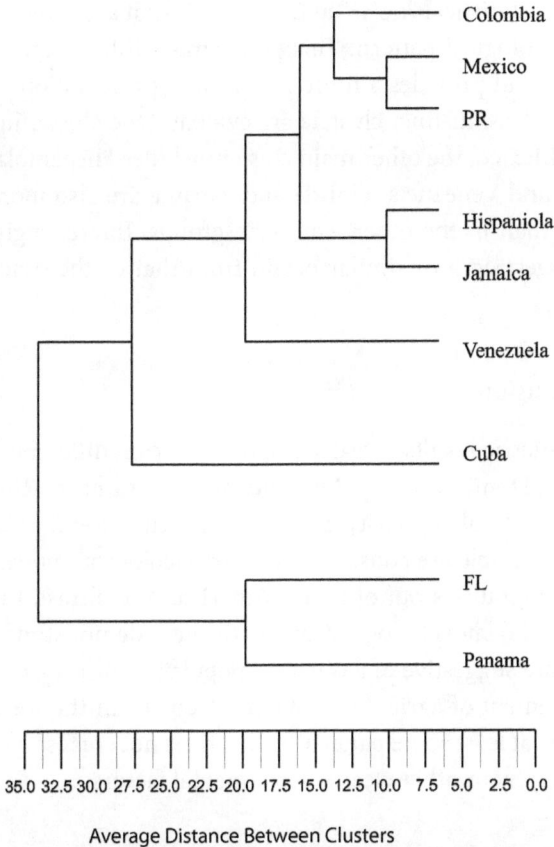

Figure 5.7. Phenogram derived from the UPGMA clustering procedure, based on the $D^2$ matrix of the first twenty PCA scores.

Average Distance Between Clusters

Table 5.9. Mahalanobis D$^2$ Distance Based on Twenty PCA Scores, Showing the Degree of Differentiation among the Groups

| | Col | Cuba | FL | Hisp | Jam | Mex | PR | Pan | Ven |
|---|---|---|---|---|---|---|---|---|---|
| Col | | <.0001 | <.0001 | .02 | .0046 | .10 | .21 | <.0001 | .20 |
| Cuba | 31.1 | | <.0001 | <.0001 | <.0001 | <.0001 | <.0001 | <.0001 | <.0001 |
| FL | 37.7 | 27.2 | | <.0001 | <.0001 | <.0001 | <.0001 | .0006 | <.0001 |
| Hisp | 12.7 | 23.1 | 42.0 | | .25 | <.0001 | .02 | <.0001 | .03 |
| Jam | 19.1 | 25.7 | 41.4 | 6.4 | | <.0001 | .49 | <.0001 | .13 |
| Mex | 8.6 | 26.4 | 30.8 | 11.2 | 14.2 | | .08 | <.0001 | .0020 |
| PR | 9.8 | 24.2 | 31.2 | 8.7 | 6.1 | 5.5 | | <.0001 | .19 |
| Pan | 28.2 | 34.0 | 16.2 | 40.3 | 35.4 | 29.4 | 24.2 | | <.0001 |
| Ven | 14.3 | 34.4 | 44.2 | 14.1 | 13.7 | 17.2 | 11.7 | 34.1 | |

Note: Upper diagonal p-values and lower diagonal distances. Italics = no significance.

remaining six principal components were excluded from the analysis, as they could be viewed as statistical "noise" and their inclusion appeared to mask the underlying biological pattern. This phenogram differs slightly from the previous tree and provides a more insightful representation of the group similarities. Three distinct clusters are evident. One cluster includes Colombia and Mexico; the other main cluster includes Hispaniola, Jamaica, Puerto Rico, and Venezuela. Florida and Panama are also more similar to each other than to the other American groups. Interestingly, Cuba is set apart, suggesting a dissimilar origin from that of the other Caribbean Taíno groups.

## Discussion and Conclusions

The traditional craniometric results reveal a lack of significant differences among the Puerto Rico, Dominican Republic, and Jamaican samples. This speaks to their morphological similarity. In addition, their general linkages to the Venezuelan sample are consistent with archaeological models pointing to ancestral migrations out of South America. In contrast, the significant differences and morphological distinctiveness demonstrated by the Cuban sample are suggestive of a different population history, one consistent with a movement of agriculturalists into Cuba from the west. All of the samples show at least some patterning that does not correspond well with simple geography or other clear environmental factors.

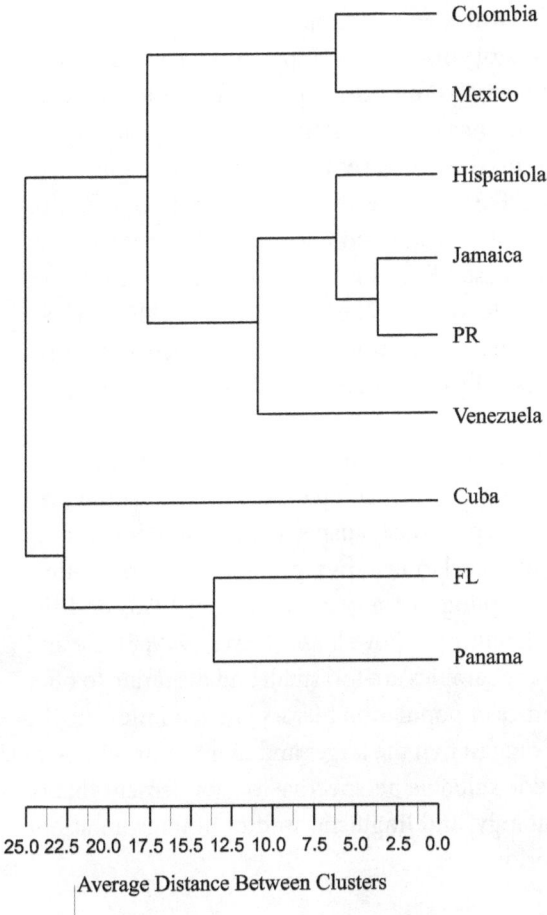

Figure 5.8. Phenogram derived from the UPGMA clustering, based on the $D^2$ matrix of the first fourteen PCA scores, showing two distinct nodes. One node comprises Colombia and Mexico, while the other cluster includes Hispaniola, Jamaica, Puerto Rico, and Venezuela. Florida and Panama are more similar to each other than Cuba, which branches away notably from the rest of the Caribbean and American series, indicating a different origin.

The geometric morphometric results are consistent with the traditional craniometric results, which partly supports the archaeological evidence suggesting a northwest dispersal from South America for the peopling of the Caribbean, evidenced by the biological similarity among samples from Colombia, Venezuela, Hispaniola, Jamaica, and Puerto Rico. However, the dissimilarity of the Cuban Taíno sample to the other Caribbean Taíno groups and to the Colombian and Venezuelan samples is suggestive of a different origin, although based on small sample sizes. The differentiation of the Cuban Taíno from the rest of the Taíno series suggests at least two separate migration routes and population sources for Caribbean settlement. One route consistent with the archaeological evidence is of a northwest movement from South America, as noted by the close biological

affinity of the Taíno groups (with the exception of Cuba) to the Venezuelan sample. A second migratory route to account for Cuban Taíno differentiation, possibly from Central America, requires further testing with
additional samples. These findings are consistent with the archaeological
evidence of Mesoamerican origins for early Caribbean settlement (Keegan 1994; Wilson et al. 1998). Furthermore, the similarity between Mexico
and Colombia suggests a similar origin and concurs with earlier studies
of possible contact between coastal Ecuador and Mexico, lending support
for likely coastal migrations (Ross et al. 2002a; Callaghan 2003b). These
results further demonstrate that population dispersal in the New World
and Caribbean was not necessarily linear in nature or based on geographic
proximity.

The patterns revealed here contribute to the growing perception of
greater heterogeneity and complexity of biological patterning within precontact Latin America than was previously suspected. The sample diversity
and nongeographic nature of the relationships are consistent with multiple
migration models for the peopling of the region (Keegan 1995) and the
Americas (Neves and Pucciarelli 1991; Powell and Neves 1999; Steele and
Powell 1993, 1994). Clearly the samples are too small and disparate to offer
a clear picture of the dynamics of population history. They do indicate the
value of this approach and suggest that the larger and more comprehensive
study under way may provide valuable perspective to complement that of
archaeology, population biology, and linguistic studies in interpretations
of the peopling of New World.

## Acknowledgments

We would like to thank Dennis Slice for generating the plots of the difference vectors and for his continued guidance in geometric morphometrics,
Beatriz Rovira for her assistance with the Panama material, Antonio Martinez for access to the Cuban collection, and Shanna Williams for generating figure 5.1. This project was funded in part by an American Museum
of Natural History Collections Study Grant and a National Museum of
Natural History Short-Term Visitor Grant.

# 6

## Crossing the Guadeloupe Passage in the Archaic Age

RICHARD T. CALLAGHAN

### Abstract

It has been suggested that two groups of Archaic peoples belonging to two different series of cultural complexes existed in the northern Lesser Antilles. One belonged to the Ortoiroid series originating on Trinidad, and the other is connected to the Casimiroid series of the Greater Antilles. Remains of either of these cultures are scarce south of the Guadeloupe Passage but not uncommon to the north. The evidence for Archaic cultures south of the Guadeloupe Passage is evaluated and computer simulations of voyaging are used to assess the difficulty in crossing it. Reasons why Archaic peoples may have avoided the southern islands and possible causes for a bias against finding sites are also explored. An alternative affiliation of the Ortoiroid series in the north is given, suggesting that these Archaic manifestations more likely belong to the Casimiroid series.

### Resumen

Ha sido sugerido que dos diferentes grupos de poblaciones del periodo Arcaico, que habitaron al norte de las Antillas Menores, pertenecieron a diferentes complejos culturales. El primero pertenecía a la serie Ortoiroid originado en Trinidad, y el segundo estaba conectado a la serie Casimiroid de las Antillas Mayores. Los restos de ambos grupos culturales son escasos al sur del Pasaje Guadalupe, pero al norte son relativamente comunes. En este trabajo se evalúa la evidencia de las culturas arcaicas al sur del Pasaje Guadalupe, y los resultados de simulaciones computarizadas de sus posibles travesías se usan para evaluar la dificultad de tales recorridos. También se exploran algunas de las razones por las cuales los grupos arcaicos pudieron haber evitado su paso hacia las islas del sur, así como las posibles causas de algún sesgo en el hallazgo de los sitios. Adicionalmente el artículo brinda una afiliación alternativa de la serie Ortoiroid del norte.

## Introduction

Rouse (1992a: 49–70) describes two early migrations of people into the Antilles: one moving west to east through some of the Greater Antilles, and another moving south to north through the Lesser Antilles. Although I will out of necessity refer to Rouse's classification system, I will use the term "Archaic" when referring to these migrations in general.

Archaic Age sites are found throughout most of the West Indies, including on Cuba, Hispaniola, Puerto Rico, the Virgin Islands, the Leeward Islands, the Windward Islands, Barbados, Tobago, and Trinidad. However, south of the Guadeloupe Passage there are remarkably few sites until one reaches Tobago and Trinidad. Tobago was separated from the mainland around 11,000 years ago, but Trinidad may have been connected to the mainland as late as 1000 BP (Kenny 1989). The possible reasons for the near absence of Archaic sites south of the Guadeloupe Passage are examined here. First, an analysis of the marine environment and the difficulty traversing the passage is given. Second, reasons (such as volcanism) why islands south of the passage might have been avoided are discussed. Third, the likelihood that Archaic sites are too deeply buried by volcanic deposits for traditional archaeological techniques to discover is considered. Finally, an alternative affiliation for the Archaic sites north of the passage in the Leeward Islands, the Virgin Islands, and Puerto Rico is suggested. These sites may belong to an Archaic expansion from Cuba and Hispaniola rather than to one originating in Trinidad.

Rouse (1992a) devised a culture-historical classification for prehistoric Caribbean cultures, and as it is the most commonly used it will be referred to here, although alternative systems have been proposed (Chanlatte Baik 1986; Chanlatte Baik and Narganes Stord 1986; Veloz Maggiollo and Vega 1982). Rouse's classification was originally developed based on ceramics (Rouse 1992a: 33) and was then extended to include Archaic traits. Curet (2003: 11–15) discusses the development of Rouse's system and presents definitions of the terminology (Curet 2003: 12). The system is hierarchical with series at the highest level. A series "consists of a set of styles or complexes that have developed one from another." A subseries consists of "smaller geographical, chronological and cultural units, intermediary between series and styles." A style consists of "all the pottery [or diagnostic artifacts] found within each people's spatial and temporal line; in practice, styles represent both the ceramic assemblages [or other artifact assemblages] and the people that created them." Series names are identified by the suffix -*oid*, while subseries are identified by the suffix -*an* (Rouse 1992a: 33). Curet (2003: 12) also notes that Rouse's (1992a) usage of the terms extends to migration routes, peoples, communities, and archaeological cultures.

The first series of interest is the Casimiroid, dating to 4000–400 BC (Rouse 1992a: 51–69). The series includes both the Lithic Age and the Archaic Age. Many authors have attributed the origin of these groups to the Yucatán Peninsula (Rouse 1992a: 51; Wilson et al. 1998). However, DNA studies on human skeletal material (Lalueza-Fox et al. 2003) suggest an origin primarily in Venezuela with a possible contribution from the southeastern United States. The sample excludes the Yucatán Peninsula. This is supported by the analysis of the problem from the point of view of navigation (Callaghan 2003c), although the results of the latter study suggest possible secondary inputs from the Yucatán Peninsula and the southeastern United States. If that is the case, though, it is most likely that people from the Greater Antilles made the initial contact. Rouse (1992a: 54) considers the series to be limited to Cuba, Hispaniola, and possibly Puerto Rico.

The Courian subseries of the Casimiroid series found on Hispaniola belongs to the Archaic Age and begins shortly after 3000 BC; it is described by Rouse (1992a: 57–60). The lithic technology is based on the preparation of prismatic cores used to produce blades. The blades are unifacially flaked around the edges to form what Rouse (1992a: 58) calls stemmed spearheads and backed blades. Ground stone is common in the artifact assemblages of cultures making up the subseries. The most distinctive of the ground stone artifacts are rectangular hammer grinders, conical pestles, and single- and double-bitted axes. There are also manos and metates and mortars and pestles.

The second series of interest is the Ortoiroid series. According to Rouse (1992a: 68–70), this series originated on the island of Trinidad and is related to cultures on the mainland both east and west of the Orinoco Delta. In the Greater and Lesser Antilles, the series spans the period 2000–400 BC.

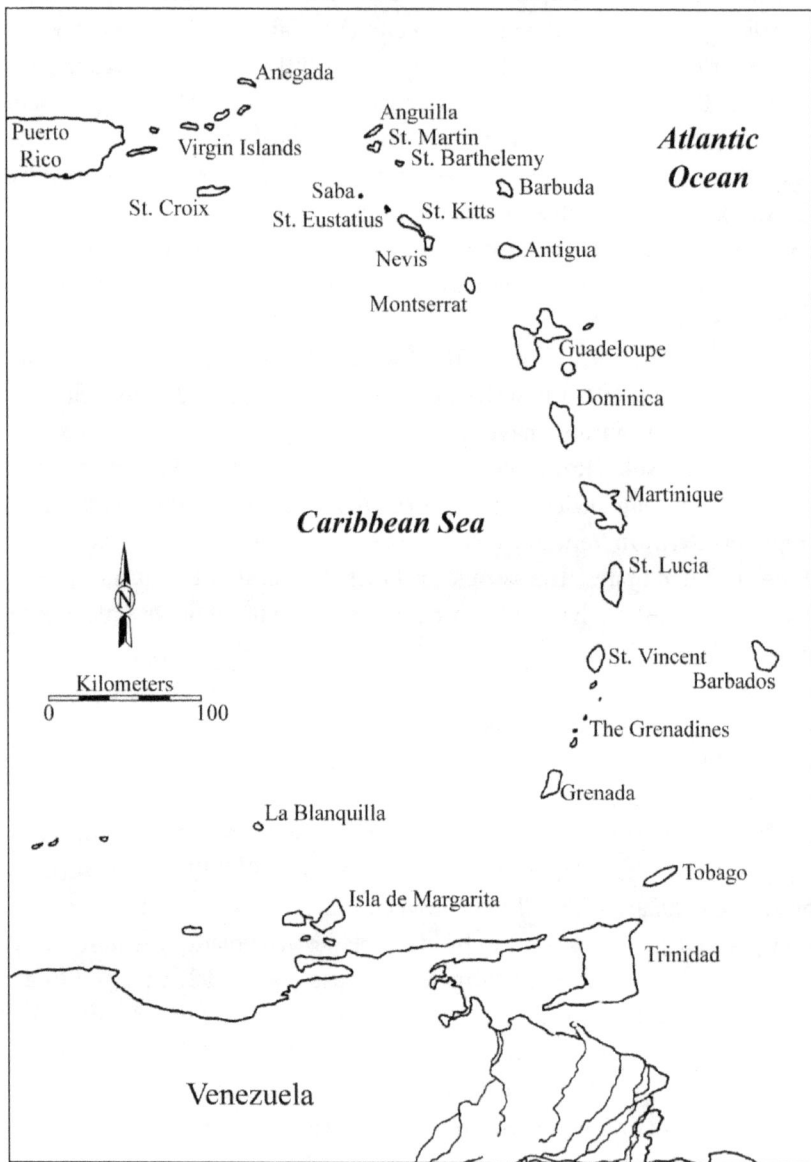

Figure 6.1. Map of the eastern Caribbean.

On Trinidad, however, this series dates back to around 6000 BC (Boomert 2000: 54). North of the Guadeloupe Passage, separating Guadeloupe from Antigua and Montserrat (figure 6.1), sites belonging to this series are most common in Puerto Rico and the Virgin Islands (Rouse 1992a: 63).

The most distinctive artifact recoveries from Trinidad are bone projectile points and barbs (Rouse 1992a: 63). Conical pestles, manos with irregularly shaped metates, simple stone choppers, and mortars are also found. Chipped stonework consists of irregular flakes. Stone net sinkers and hammerstones are included in the assemblage on Trinidad. Boomert (2000: 74) states that the edge grinders in the assemblage "should be considered as type artifacts of the Ortoiroid series." On other islands to the north, assemblages include edge grinders and stone and shell celts, but these tool types are not necessarily on all islands where the Ortoiroid has been identified. Rouse (1992a: 63) notes that "the series is difficult to define and to divide into subseries because its sites contain small numbers of artifacts, all simple types with so few traces of manufacture that it is hard to define diagnostic traits."

For the Corosan subseries of Puerto Rico (Rouse 1992a: 65–66), edge grinders are diagnostic and sites also contain pendants and beads made from shell, bone, and stone. Aside from the above, there are nonmanufactured objects of shell, stone, and bone that show use as cutting edges, grinders, choppers, hammerstones, and picks. Some sites are reported to have chipped stone tools, and Rouse notes that conical pestles and one stone bowl have been found. He attributes these last artifacts to contact with Casimiroid peoples. Corosan sites in general consist of groups of small piles of shell and artifacts.

The main focus here, however, is on the Lesser Antilles and Virgin Islands. North of the Guadeloupe Passage and outside of the Greater Antilles, Archaic sites are found on St. Thomas (Lundberg 1989); Anguilla, including the small offshore Dog Island (Crock et al. 1995; Crock and Petersen 1999); Saba (Hofman and Hoogland 2003; Hofman et al. 2006a); St. Martin (Knippenberg 1995); St. Kitts (Armstrong 1980); Nevis (Wilson 1989); Barbuda (Watters et al. 1992); and Antigua (Davis 1974, 1993, 2000; de Mille 2005; Nicholson 1976; Olsen 1973). South of the Guadeloupe Passage, far fewer Archaic sites have been reported, including ones on Guadeloupe (Richard 1994), Martinique (Allaire and Mattioni 1983), St. Vincent (Hackenberger 1991), and Barbados (Drewett 1995).

The sites north of the Guadeloupe Passage are, for the most part, considered to be related to the Ortoiroid series, and there is a fairly high level of variability in the associated assemblages. For St. Thomas, the Krum Bay culture (Lundberg 1989) is very similar in both site content and locations to the Corosan of Puerto Rico. They tend to be small accumulations of shell and artifacts, typically on the coast. The main difference is the inclusion in the assemblage of chipped and sometimes partially ground stone celts. Chipped stonework made expedient use of resources with little concern for the form of the tool (Lundberg 1989: 119).

On Anguilla there are five Archaic sites, including one on adjacent Dog Island (Crock et al. 1995; Crock and Petersen 1999). Taken together, the assemblages include shell tools and vessels, chipped and ground stone tools, utilized coral, and shell and stone debitage. There are rarely stone axes, pendants, abraders, and hammerstones. Other rare, but interesting, finds are blades or bladelike flakes, one of which has been compared to blades found in the Dominican Republic (Crock and Petersen 1999: 139). At the Whitehead's Bluff site (Crock and Petersen 1999), imported Antiguan chert was used in an opportunistic way to produce flakes. There appears to have been a high degree of conservation of the material.

The single known Archaic site on Saba, Plum Piece (Hofman and Hoogland 2003; Hofman et al. 2006a), also shows that the main lithic material used was from Antigua. Similar to chipped stone on Anguilla, an expedient flake technology was used. Some bladelike flakes, but no blade cores, were found. On St. Martin, the Norman Estate site (Knippenberg 1995) contains short-term Archaic occupations. Here much of the stonework utilizes Antiguan chert, but local materials were also used. While both materials were used to produce small flakes in a nonsystematic way, the Antiguan chert did not undergo the same reduction sequence as the local material. This may indicate an attempt to more fully utilize the exotic resource. St. Kitts (Armstrong 1980) has two Archaic Age sites containing shell celts and tools along with a few chert flakes. Wilson (1989) reported two Archaic Age sites from Nevis. Only faunal remains are noted, but the absence of ground stone tools is stated. An additional Archaic site is reported on Barbuda (Watters et al. 1992). Shell and stone are reported, but only a shell celt is described.

Recent work by de Mille (2005) focuses on the Archaic assemblages of Antigua and is undoubtedly the most thorough treatment of the subject to date. In comparing the ground stone artifacts from Antigua with those of other islands in the Caribbean, de Mille (2005: 330–37) notes that ground stone is not always part of other assemblages. Further, much of it is shaped by use rather than made into formal shapes. Antigua has a greater abundance of ground stone that is made into a variety of formal shapes; this pattern appears more closely related to that found at Casimiroid sites in the Greater Antilles.

In analyzing the chipped stone assemblage from Antiguan Archaic sites, de Mille (2005: 353–61) notes that production of blades and flakes is structured but that the nature of the tools themselves is expedient: that is, there is no formal tool kit, but the technology for producing blades is formal. The production of blade cores and blades is likely a factor constrained by the quality and availability of lithic resources. De Mille (2005: 361) points to "the almost perfect correlation of blade production with suitable raw material sources." She further points to the behavioral similarities between the pattern of choice and use of blades and flakes made by nonblade techniques. It appears that when high-quality material was available, blade-producing technologies were implemented, but otherwise a more conservative, and expedient, strategy was employed.

Rouse (1992a: 68) considered there to be an influence from cultures belonging to the Casimiroid series on Antigua. This influence is manifested in the Antiguan lithic industry, which includes blades unlike those of the cultures relegated to the Ortoiroid series. Davis (2000: 99) suggests, without elaborating on the nature of the relationship, that the Archaic lithic industries of Antigua are related to the Casimiroid series.

South of the Guadeloupe Passage, one small site reported by Richard (1994) contained shell and lithic tools, but no description is given. Two small sites on Martinique are considered to belong to the Ortoiroid series (Allaire and Mattioni 1983). The small assemblages included edge grinders, flakes, and an ax fragment. Hackenberger (1991) reported an Archaic site on St. Vincent with lithic flakes, while on Barbados (Drewett 1995) a single radiocarbon-dated adze is difficult to classify on its own.

The paucity of sites south of the Guadeloupe Passage is puzzling. Several environmental factors need to be examined to determine the likelihood that this is simply a sampling bias reflecting less research in the south. The following sections evaluate the difficulty of navigating the passage, whether volcanic activity could have acted as a deterrent to settlement, and whether volcanic activity may have obscured sites.

## Computer Simulations of Navigating the Guadeloupe Passage

According to Clarke's (1989: 44) summary of weather patterns in the area today (figure 6.2), the Caribbean lies within the wind belt known as the Northeast Trades. With the exception of disturbances from tropical cyclones, the weather is quite stable. The prevailing winds are easterly and usually steadiest in the south of the region between December and May. Summer and fall are warmer and more humid than winter and spring. Cloud cover and rainfall increase, as does thunderstorm activity, and winds are often lighter and more variable in the spring.

Tropical cyclones occur most frequently in summer and fall. The northern limit of the Northeast Trades is 28°N and is reached between July and September. At this time, the strongest and steadiest winds pass through the middle of the region; near the northern limit, they tend to be more variable. The limit shifts south to about 24°N between February and April. On average, the winds blow at 11–15 knots from the east-northeast.

Winds in the passage are predominantly from the northeast and southeast at speeds generally between 11 knots and 15 knots throughout the year. A westward-setting current runs between Antigua and Guadeloupe, while the current between Guadeloupe and Montserrat sets to the northwest (Defense Mapping Agency Hydrographic/Topographic Center [DMAH/TC] 1985: 144). The currents often reach speeds of 3 knots (Stone and Hays 1991: 224). Climate change is a consideration here, although I have discussed this in some detail elsewhere (see Callaghan 2003c: 327–29). The conclusion was that while there were probably differences in current speed, these were not highly significant to require changes in the parameters.

Previous work (Callaghan 2001, 2003c) analyzing the difficulty of traveling by canoe through the islands of the Caribbean and maintaining contact with the South American mainland during the Archaic and Ceramic ages suggests that there were no environmental or technological barriers. However, these studies investigated routes to the west of the Lesser Antilles and did not consider individual passages between islands. The same computer program (Callaghan 2001, 2003c) in these studies is used here to investigate traversing the Guadeloupe Passage in both directions: between Montserrat and Guadeloupe and between Antigua and Guadeloupe.

The dugout canoe design evaluated in this study is based on the Stargate canoe, recovered in the Bahamas by Stephanie Schwabe. This watercraft is remarkably similar to canoes still being used by the Ye'Kwana and other

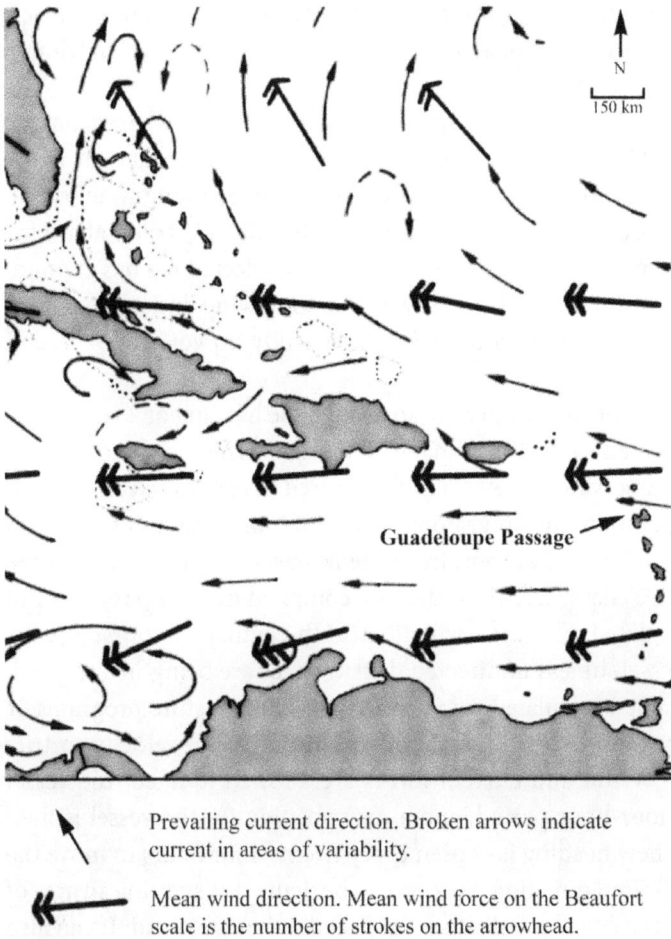

Figure 6.2. Caribbean winds and currents.

native groups on the upper Orinoco River of Venezuela today (Callaghan and Schwabe 2001). Canoes of this style were analyzed in Venezuela to determine their performance characteristics, including speed, leeway, carrying capacity, and stability. These characteristics were also analyzed using naval architecture programs. The size of the canoe used here was 8.33 meters. Speed is a major consideration in this study; it was determined that this type of canoe could be propelled by four crew members at a maximum of 3.4 knots. It was also assumed in the simulation that the entire crew was made up of eight people who could paddle in shifts. This is the same assumption used in previous work (Callaghan 2001, 2003c). However, given the short distances involved in crossing the Guadeloupe Passage, it is of less importance.

The wind and current data used in this program are based on the United States Navy's *Marine Climatic Atlas* (1995). The data set is organized in one degree Marsden squares (one degree of longitude by one degree of latitude). The program automatically shifts to the database for the following month after the month originally selected for has expired. This feature better reflects the reality of changing wind and current conditions over long voyages, although in this study no voyages exceeded twenty-four hours.

The program allows the operator to change the heading of a vessel during a voyage to reflect decisions made by the crew. This last feature is important when assessing the level of skill required to reach a selected target. In its basic operation, the program makes a random selection of direction and speed for wind and current from the *Marine Climatic Atlas* (United States Navy 1995) database. These data are compiled from ship reports and other sources dated after the early nineteenth century. A course is chosen for the vessel, unless undirected drift voyages are being investigated. Performance data, calculated using either naval architecture programs or field tests, are then used to calculate the ratio of vessel velocity to true wind velocity. Wind and current forces are allowed to affect the vessel for a twenty-four-hour period, and a new position for the vessel is then calculated. A new heading is chosen every twenty-four hours to move the vessel in the desired direction. Success can be defined as sighting an area of interest such as a coastline of either the mainland or an island. It can also be defined as actually making landfall.

The results of the simulation suggest that there would have been little difficulty for Archaic Age peoples in navigating the Guadeloupe Passage in either direction from Montserrat, Antigua, or Guadeloupe. Even if paddled speeds of only 2 knots were maintained, these short voyages would mostly be successful. The few unsuccessful voyages occurring in midwinter and midsummer only result in returning to the island of origin. Traveling from south to north does appear to be slightly more difficult but is in no way difficult enough to account for the observed settlement patterns and is a seasonal problem. Despite the relative ease of making the crossing, there are some environmental conditions to consider. The deep valleys and steep, high topography of Guadeloupe can produce heavy and sudden squalls that are a danger to travelers (DMAH/TC 1985: 152). Given the results, the reasons for the paucity of sites south of the passage are unlikely to be related to oceanographic conditions or technological deficiencies.

## Volcanism as a Deterrent to Settlement

There may be other reasons for the observed settlement patterns. If we accept the designation of sites north of the Guadeloupe Passage as belonging to the Ortoiroid series, then we might expect more sites in the Windward Islands as the series originates in Trinidad (Boomert 2000). However, as I have pointed out elsewhere (Callaghan 2001), and as the above simulation suggests, there were no environmental or technological factors requiring migrating peoples to stop at each island along the way. But why would they not choose to do so?

Davis (2000) has suggested a correlation between Archaic site locations on Antigua and shallow marine and chert resources. The islands south of the Guadeloupe Passage are primarily high volcanic islands that rise steeply from the ocean. Shorelines with a low gradient are perhaps more limited than in the northern islands, but they do exist. Furthermore, the Grenadines between St. Vincent and Grenada have ample environments of this nature. The inland locations of sites on Saba (Hofman and Hoogland 2003) and Martinique (Allaire and Mattioni 1983) also suggest that settling next to shallow marine resources was not always the primary factor. The absence of high-quality chert resources on islands south of the passage may be a factor, but most of the islands to the north present the same situation.

The Lesser Antilles is a volcanically active island chain, and this could possibly have discouraged settlement in the Archaic age, especially considering that several of the volcanoes have been known historically to erupt almost simultaneously (Lindsay et al. 2005a: xv). During the period between approximately 5000 BC and AD 1, there are indications that volcanoes were active in the region, but not everywhere (Lindsay et al. 2005a). It is worth examining patterns of volcanic activity.

Starting in the south of the Windward Islands, the first active volcano is Mount St. Catherine on Grenada. This volcano has not erupted in historic times (Robertson 2005a: 50). However, there was some slight activity reported in 1867 and 1902 (Robertson 2005a: 57). There are also explosion craters that may be less than one thousand years old (Robertson 2005a:

55). Evidence for eruptions during the period of interest here does not currently exist. Eight kilometers to the north is the submarine volcano Kick 'em Jenny, and five kilometers farther to the east is the volcanic center of Ile de Caille. Kick 'em Jenny, with its summit measuring 180 meters below the surface (2003 observations; Lindsay and Shepard 2005: 115), is the only submarine volcano in the region to erupt in the past five hundred years; it has erupted more times historically than any other volcano in the Lesser Antilles. There have been at least twelve eruptions since 1939 (Lindsay and Shepard 2005: 108). The average is one eruption every five years (Lindsay and Shepard 2005: 118). Kick 'em Jenny is rising from the ocean out of a much larger crater, Proto–Kick 'em Jenny (Lindsay and Shepard 2005: 116). Kick 'em Jenny can be a serious hazard to seafarers—in 1939, the volcano ejected material 300 meters above the sea's surface (Lindsay and Shepard 2005: 112. Ile de Caille (Lindsay and Shepard 2005: 122) is the youngest volcano in the Lesser Antilles and may be as young as one thousand years old. Caliviny pottery dating to around 800–1,200 years ago has been reported embedded in the lava. As with Mount St. Catherine on Grenada, there are currently no dates for eruptions during the Archaic period.

The Soufrière volcano on St. Vincent has been characterized by explosive magmatic eruptions (molten rock), ash falls, pyroclastic flows (ash and pumic) and surges (highly fluid, fast-moving rock), and lahars (mud flows) (Robertson 2005b: 241). Historically, Soufrière has had both quiet effusive and explosive eruptions with one catastrophic, violently explosive magmatic eruption, or plinian event. Eruptions (Robertson 2005b: 248) were recorded in 1718, 1780, 1812–14, 1902–03, 1971–72, and 1979. Smaller events were reported between these dates and are currently ongoing. Mudflow deposits are in some instances 25 meters thick (Robertson 2005b: 246–48). The 1718 explosive eruption dropped ash on Martinique, St. Kitts, Barbados, and Hispaniola. The sound was heard in Antigua to the north and Trinidad to the south. Over the past four thousand years, there has been on average one explosive eruption every one hundred years. Dates for volcanic rocks on St. Vincent from two thousand to five thousand years ago suggest considerable activity (Robertson 2005b: 249–50).

St. Lucia has a number of volcanic centers (Lindsay 2005: 219). Historic activity has been limited to phreatic (steam) eruptions, the last of which occurred in 1766 (Lindsay 2005: 226). Available dates for prehistoric eruptions begin several million years ago but end about twenty thousand years ago. This does not necessarily mean that St. Lucia's volcanic centers were inactive during the intervening period, as a number of explosion craters and domes are younger than twenty thousand years old (Lindsay 2005: 237). Earthquake activity historically has been recorded fairly frequently and in the very recent past.

Martinique's Montagne Pelée (Boudon et al. 2005) is probably the most famous volcano in the Caribbean, because of the 1902 eruption that killed thirty thousand people. Historically (Boudon et al. 2005: 137), there have been eruptions in 1792, 1851, 1902–05, and 1929–32, as well as eighteen magmatic eruptions, made up of molten rock, gases and crystals, in the past five thousand years, with thirteen occurring between 5000 BC and AD 1. Several of these eruptions have been plinian, the last occurring as recently as 650 years ago. There were also an unknown number of phreatic eruptions; it is likely that the actual number of eruptions may be largely underestimated.

Nine volcanic centers are found on the island of Dominica (Lindsay et al. 2005b: 2). The most recent magmatic eruption was about five hundred years ago, and two phreatic eruptions were recorded in 1880 and 1997. At least four eruptions are indicated between AD 1 and 5000 BC. One of these seems to have lasted for a fairly long time, as indicated by a series of overlapping radiocarbon dates. Besides volcanic activity, seismic activity is relatively frequent, with a number of earthquake swarms recorded historically.

Within the Guadeloupe group, Basse Terre has the only volcanic centers (Komorowski et al. 2005: 68). A number of historical phreatic explosive eruptions have occurred: in 1690, 1797–98, 1812, 1836–37, 1956, and 1976–77 (2005: 80). In 1440, there was a magmatic eruption, followed by a lahar flow in 1550, a pyroclastic surge in 1590, and a scoria fall in 1600 (Komorowski et al. 2005: 80). Considerable outgassing and earthquakes have also occurred (Komorowski et al. 2005: 79). Between AD 1 and 5000 BC, twenty-four separate events are indicated by pyroclastic deposits alone (Komorowski et al. 2005: 79).

North of the Guadeloupe Passage, five islands are considered volcanically active—Montserrat, Nevis, St. Kitts, St. Eustatius, and Saba. Montserrat has been erupting since 1995; this is the first eruption since European settlement in 1632 (Hincks et al. 2005: 148). Pyroclastic flows started about four thousand years ago, after a long period of inactivity. The last eruption before 1995 was probably just prior to European settlement (Hincks et al. 2005: 153). Nevis (Simpson 2005: 170) has not had an eruption recorded in historical times, and there is no evidence of eruptions during the time that humans occupied the island. St. Kitts (Robertson 2005c: 207) has not erupted historically. Just prior to AD 1, there are four radiocarbon dates that overlap for volcanic rocks, and there are two more overlapping dates about three hundred years earlier. There are three other dates: around 1000 BC, 1600 BC, and 2200 BC. On St. Eustatius (Smith and Roobol 2005b: 193), the last eruption was sometime between 1,635 and 1,755 years ago. Other evidence points to volcanic activity before 5500 BC, but none recorded in the period of interest. Saba is the last island with volcanic activity in the chain of the Lesser Antilles, with the last eruption occurring about 280 years ago, just before European settlement (Smith and Roobol 2005a: 181). There is evidence that humans may have occupied the island while it was having brief periods of volcanic activity around 800 BC (Smith and Roobol 2005a: 184).

From the above, it is clear that there was more volcanic activity south of the Guadeloupe Passage than to the north during Archaic times. In the north, there were Archaic occupations of the active volcanic islands of Nevis, St. Kitts, and Saba. Dates for occupation of Nevis are from around 600 BC (Wilson 1989: 435). As noted above, there is no evidence of volcanic activity near the time of human occupation. The Archaic settlements on St. Kitts (Armstrong 1980) date from about 4,100 to 2,175 years ago. The later site appears to have been buried by volcanic deposits shortly after the occupation. On Saba, the Archaic occupation is dated to between 1520 and 1875 BC (Hofman and Hoogland 2003: 16), a period for which we have no evidence for volcanic activity.

To the south, the Archaic site on Guadeloupe, Grand Terre (Richard 1994), is from a nonvolcanic formation dating to near the end of the first millennium BC. Activity that seems to have occurred around that time is evidenced by debris avalanches and lahars. Given that the site is near St. François, it is about the farthest location from volcanic centers as was possible. On Martinique (Allaire and Mattioni 1983), the single date for two Archaic sites are quite late, from AD 260 to 440. Interestingly, this falls well into the Saladoid period, and some activity is indicated around that time. The Archaic material on St. Vincent (Hackenberger 1991) is undated and is considered by many to be dubious (Keegan 1994: 266), a conclusion with which I concur.

From this evidence, then, it is possible to conclude that Archaic peoples avoided the more actively volcanic islands. The single shell celt dating to

1630 BC on Barbados (Drewett 1995), a nonvolcanic island, is suggestive of an early settlement during the Archaic Age, but a more well-developed chronology and additional artifactual evidence are required to determine the age and extent of settlement during this time. Volcanism does not fully explain the absence of sites in the Grenadines and some of the smaller nonvolcanic islands of the Windward group. Volcanism does not seem to have greatly discouraged settlement on these islands by later peoples, although islands may have been temporarily abandoned during extreme volcanic and seismic activity.

## Volcanism as a Sampling Bias

A last consideration regarding volcanism is its possible effect on finding early sites. Volcanic eruptions may in some cases have buried Archaic sites with such deep deposits that they would be virtually unobservable and/ or inaccessible to archaeologists. The volcanoes of St. Vincent, Dominica, and Martinique are discussed by Sigurdsson and Carey (1991). For St. Vincent, the main formation that corresponds to the Archaic period is the Yellow Tuff Formation (Sigurdsson and Carey 1991: 71–74). The depth of the formation is up to twenty meters and is mainly scoria composed of basaltic or andesitic materials, and coarse ash with a small amount of lithic fragments.

Associated with the Yellow Tuffs are well-bedded pyroclastic falls. Near Sandy Bay on the northeastern tip of the island, these pyroclastic fall deposits reach depths of up to fifty meters. The deposits from this formation can be found all the way to the south coast to Kingstown, where it is two meters thick. The age of the formation is suggested to be around 3700 BC, with radiocarbon dates ranging from 2900 to 2000 BC. There are indications that there were periods of dormancy in the form of paleosols. However, these are few in number, and a lack of charcoal in most of the horizons indicates that no significant vegetation was able to develop during the period of formation. Overlaying deposits from later activity has been dated from about AD 700 to the present. There may not be a dormant period between these deposits and the Yellow Tuff Formation. It is possible that lava flow eruption was the dominant type of eruption at the time.

During a recent archaeological survey on St. Vincent (Callaghan 2007), apparent single-event ash layers were observed at least ten meters deep. Certainly this would impede the identification of early sites. Insular

Saladoid sites (circa 200 BC) were recorded beneath ash fall and alluvium at depths of up to three meters. The sites were located because of fortuitous exposure. However, this kind of ash fall is not found everywhere, and a sampling bias due to volcanic activity cannot explain the absence of later Archaic Age sites on St. Vincent or in the Grenadines.

Sigurdsson and Carey (1991) provide no information regarding the extent or depth of deposits for eruptions during the Archaic period for Dominica. However, Lindsay and colleagues (2005b: 21–38) evaluated the hazards posed by the active centers on Dominica; from this, some idea of relatively recent past effects can be suggested. From the Morne Canot vent near the southwest tip of the island, ash falls are estimated to occur up to 9.5 kilometers away. Only within 1–2 kilometers would the deposits be of significant depth, approximately 30 centimeters. Very similar effects would be seen given an eruption at the nearby vent at Morne Patates. Also nearby, an eruption from Morne Anglais would only have ash fall of about 50 centimeters in an area 1–5 kilometers away from the center of the event. The maximum extent of ash fall would cover the southern third of the island but at 11 kilometers would only reach a depth of 5 centimeters. An eruption at the volcanic center of Micotrin in the south-central part of the island would have ash deposits of up to 30 centimeters in the immediate vicinity. Although deposits would cover the southern third of the island, for the most part, depths would only be 1–5 centimeters. The most dangerous scenario is an eruption from the Wotten Waven caldera in the south-central part of the island. The hazard would extend to the southern half of the island, but ash falls are estimated to be similar to Morne Anglais. At the northern tip of Dominica, the Morne aux Diables center would have ash fall depths of 1–30 centimeters. Although this may be a problem for occupants within the hazard zone, it is not a depth that would greatly obscure archaeological remains.

Over the past 5,000 years, six major pumice-forming eruptions occurred on Martinique (Sigurdsson and Carey 1991: 12–15). On average, eruptions of this type occur every 750 years. These eruptions can be classified into three common types. The first type consists of a thick pumacious pyroclastic flow. The flow is then overlain by a thick deposit of fine ash. The second type starts with plinian pumice fall mixed with lithic-rich ash. These deposits form a thin layer covered by a pumice-rich flow topped with a thin layer of ash. The third type is comparable to the second, but there are surge deposits interbedded between pumice-rich pyroclastic deposits. These types of deposits are found over the northwestern part of Martinique. Pyroclastic flows associated with these events are largely confined to the valleys of the southwestern and northeastern flanks of Montagne Pelée. The thickness of the flows may be as much as 50 meters. Other types of deposits are found outside of the valleys. These are ash hurricane falls and ash cloud falls. The former type results from a low-density cloud with a great deal of turbulence. The latter type consists of low fine-grained materials. The deposits can extend as much as 10 kilometers from the source and may occasionally reach depths of over 4 meters. Although more active recently than Dominica, Martinique does not appear to have had extensive deposits that could have deeply buried Archaic sites over most of the island.

Volcanic activity south of the Guadeloupe Passage does appear to be greater than to the north. Deposits on St. Vincent may have blanketed the entire island, but not for the entire Archaic Age. On other volcanic islands such as Dominica and Martinique, volcanic deposits that could have deeply buried Archaic Age sites do not cover the whole of the islands. Further, there are several areas within the Guadeloupe group, Barbados, and the Grenadines where volcanism cannot explain the absence of Archaic sites. Some other explanation for their paucity south of the Guadeloupe Passage is necessary.

## Discussion

A number of avenues have been examined in an attempt to account for the near absence of Archaic Age sites in the Lesser Antilles south of the Guadeloupe Passage. Simulations of voyaging do not suggest any obstacles due to either technological limitations or the marine environment. Simple dugouts using human power could have made the trip in both directions. Furthermore, later Saladoid peoples clearly had no problem crossing the passage.

The second avenue of investigation considers differences in the levels of volcanic activity north and south of the passage during the Archaic Age. Certainly the southern islands demonstrate a higher level of activity during the period than in the north. However, not all islands in the south are volcanically active, and for some, there is little evidence of prolonged volcanism. Archaic peoples could have safely settled on the islands of the Grenadines, Barbados, and Grand Terre and Marie-Galante in the

Guadeloupe group. The single small site on Grand Terre suggests that this occurred. The two small, and curiously late, sites on Martinique may also support an Archaic occupation south of the Guadeloupe Passage. With the exception of a single [14]C date from an Archaic-type shell celt from Barbados, no evidence currently exists for peoples settling the island prior to the Saladoid period.

It is possible that belief systems and fear kept peoples of the Ortoiroid series from settling in the southern islands and peoples of the Casimiroid series from moving in from the northwest. This premise is not currently testable with archaeological techniques. It would suggest that the Ortoiroid peoples bypassed the southern islands and settled in the north. The available radiocarbon dates and previous seafaring simulations, if taken at face value, indicate that the later Saladoid peoples from the Orinoco region of South America settled in the north first and then moved south (Callaghan 2001; Fitzpatrick 2006). However, this pattern could simply be an artifact of few radiocarbon dates and a lack of archaeological investigations in the southern islands.

The final avenue of investigation is volcanism as a sampling bias obscuring Archaic sites. While St. Vincent may fit into this scenario, Dominica does not. Even the more volcanically active Martinique does not seem to have deposits mantling the entire island to depths that would make finding Archaic sites impossible. Davis (2000) suggested that Archaic sites are associated, in part, with shallow marine resources. This preference may mean that Archaic Age sites are particularly affected by coastal erosion (see Kaye et al. 2005; Fitzpatrick et al. 2006). However, there are the sites on Saba and Martinique that are not susceptible to shoreline erosion. Finally, locations such as the Grenadines noted above would not have great depths of ash fall, if any.

There is another possible explanation, but one that will need considerable work to evaluate. This explanation is related in some ways to the second avenue of investigation presented here. It may be that the peoples of the Casimiroid series were at their frontier on Antigua and other islands in the north when the Saladoid people arrived, possibly as early as 500 BC. As for peoples of the Ortoiroid series, it may be that they did not venture farther north than Tobago or did so only in an exploratory manner. Even on Tobago, aside from a number of isolated finds that appear to be Archaic, there is only one site of this age that dates to the third millennium BC (Boomert 2000: 77).

If we look at known Archaic sites to the north of the Guadeloupe Passage, fifty are on the island of Antigua. This is several times the number of Archaic sites on all other islands outside of the Greater Antilles put together. The material from Antigua classified by Rouse into the Ortoiroid series has also been associated with the Casimiroid by both Rouse (1992a) and Davis (2000). Several Ortoiroid sites in the northern Lesser Antillean islands also contain materials that fit within the Casimiroid series. Ortoiroid series sites in the Greater Antilles belong to the Corosan subseries and are limited to Puerto Rico. Rouse, however, notes artifacts that would fit within the Casimiroid series in Corosan sites.

The issue arises, then, of the proposed distinctiveness of Ortoiroid assemblages north of the Guadeloupe Passage. As noted previously, Boomert (2000: 74) considers the type artifact for the series to be the edge grinders. Despite this being the type artifact, he points out that the artifact has a distribution as far west as Panama during the Archaic period. Rodríguez Ramos (2005: 1) states that "[o]ne of the most ubiquitous artifacts in archaeological contexts in Puerto Rico and the rest of the Caribbean is the edge-ground cobble (also known as edge-grinders, pebble grinders, edged cobbles, *pollissoir latreaux, manos simples* or *majadores laterals*)." He further points out (Rodríguez Ramos 2005: 4) that these artifacts are found not only in Archaic Age sites but also in Saladoid Ceramic period sites and later Ostionoid sites and as far along the north coast of South America as Panama. Similar artifacts are found outside of the circum-Caribbean in North America as far away as California (Rodríguez Ramos 2005: 13).

Given the broad chronological and geographical distribution, how can edge grinders be viewed as diagnostic or type artifacts of the Ortoiroid series? Certainly they cannot in their own right. For instance, given the date for the sites of Boutbois and Le Godinot on Martinique (Allaire and Mattioni 1983), it is more probable that the sites are aceramic Saladoid sites versus Ortoiroid sites. Without these two sites, there is only the site near St. François on Grand Terre, Guadeloupe (Richard 1994), and the celt from Barbados that might be considered Archaic with even a modest degree of confidence (see Fitzpatrick 2006 for a discussion on the reliability and usefulness of single radiocarbon dates). The barbed points that Rouse (1992a: 63) considered distinctive of the Ortoiroid series on Trinidad are not found north of that location. The differences between the blade technologies and expedient flake technologies north of the Guadeloupe Passage (de Mille 2005: 361) do not translate into behavioral differences and may be simply due to a lack of suitable raw material. Most Ortoiroid sites to the north of the passage are small and short-term occupations, although people may have returned to them fairly regularly. Exceptions to this exist on Antigua, where lithic resources suitable for a blade technology are plentiful. Finally, it is interesting that when other materials that are considered diagnostic are found in the north, they indicate associations to the Casimiroid series.

Keegan and Diamond (1987) felt that the evidence at the time did not support the notion of small islands in the Lesser Antilles being permanently settled by Archaic Age hunter-gatherer-fishers. Twenty years later, the evidence has not changed, although Antigua may be the exception. Takamiya (2006) reviews some of the reasons why people with an Archaic type of subsistence do not permanently settle small islands except under certain conditions. For his case study on Okinawa, the key to allowing settlement appears to be the availability of storable nuts, which balanced a protein-rich diet based on marine resources. It would be interesting to identify carbohydrate resources on Antigua in Archaic sites that could have filled similar dietary requirements. The options here would be a wild plant not found elsewhere in the Lesser Antilles or something from the Greater Antilles brought to Antigua by Archaic Age peoples.

In summary, the current evidence for an Ortoiroid migration beyond Tobago is very ephemeral. What has been classified as Ortoiroid could at least as easily be Casimiroid. An examination of volcanic activity does not indicate that these events could be responsible for biasing the sample by burying sites too deeply for detection. Does this mean that there was no Ortoiroid migration? Perhaps not, but more work is required to establish whether one occurred or not. Armstrong (1980) noted the difficulty of finding Archaic sites on volcanic islands. Since his initial observations, much more information has become available to archaeologists regarding the extent of volcanic deposits that can be used to design appropriate survey strategies, especially on nonvolcanic islands. Equally important would be the undertaking of a detailed analysis of Archaic materials north of the Guadeloupe Passage with an in-depth comparison with Archaic materials on Trinidad. This would then give archaeologists a better idea of whether or not the similarities between the two areas are significant enough to postulate a direct connection and migration. In the meantime, the Ortoiroid migration should be seen as tentative.

## Acknowledgments

I would like to thank Scott Fitzpatrick and Ann Ross for inviting me to present this paper in the symposium "New Perspectives on the Prehistoric Settlement of the Caribbean" at the 2006 Society for American Archaeology conference in Puerto Rico and a further thanks to Scott Fitzpatrick for his comments on a draft of the paper. I would also like to thank Michael Turney for the figures and Alejandra Alonso for the Spanish translation of the abstract.

# 7

# Interisland Dynamics

## Evidence for Human Mobility at the Site of Anse à la Gourde, Guadeloupe

MENNO L. P. HOOGLAND, CORINNE L. HOFMAN,
AND RAPHAËL G. A. M. PANHUYSEN

## Abstract

The existence of elaborate exchange networks has been attested to by the distribution of exotic raw materials, goods, and ceramics styles throughout the Caribbean islands. Exchange, both regional and microregional, is regarded as instrumental in shaping and maintaining social relationships between islanders. The search for marriage partners is seen as one of the driving forces behind the establishment and maintenance of these relationships. Analysis of strontium isotopes on human teeth from the site of Anse à la Gourde on the island of Guadeloupe in the northern Lesser Antilles provides insights into the mobility of the island inhabitants. Results of these analyses are presented and subsequently interpreted in the wider context of mobility and exchange in the pre-Columbian insular Caribbean.

## Resumen

La existencia de cadenas de intercambio elaboradas se ha comprobado mediante la distribución de materias crudas exóticas, bienes, y estilos de cerámicas a través de las islas. El intercambio, ambos regionales y microregionales, se contempla como un elemento instrumental en la formación y el mantenimiento de relaciones sociales entre los isleños. La búsqueda de parejas matrimoniales es la fuerza que maneja el establecimiento de dichas relaciones. El análisis de isótopos de estroncio de dientes humanos del sitio Anse à la Gourde, ubicado en la isla de Guadalupe en las Antillas Menores norteñas, provee indicios en torno a la movilidad de los habitantes de la isla. Los resultados de dicho análisis son presentados e interpretados dentro del contexto más amplio de movilidad e intercambio en el Caribe insular precolombino.

## Introduction

Deciphering patterns of pre-Columbian interaction and exchange on regional and microregional scales has been an important facet of research among Caribbean scholars over the past decades. On a regional scale, the distribution of linguistic and biological aspects as well as pottery styles and other cultural remains has been used to trace the initial migrations of early Ceramic Age groups from the South American mainland into the Antilles. Moreover, these were put forward to explain the maintenance of mainland contacts during the Ceramic Age and early colonial period, the changes in social organization, the "wealth finance" of "big man collectivities" and chiefdom societies, and the sociopolitical and esoteric interaction between the Lesser and Greater Antilles during the emergence of the Taíno *cacicazgos* of the Late Ceramic Age (see, for example, Allaire 1990; Boomert 2000; Cody 1991; Crock 2000; Crock and Petersen 2004; de Waal 2006; Keegan and Maclachlan 1989; Rodríguez López 1991a, 1991b; Roe 1989, 1995). Initially, focus was laid on long-distance migrations and population movement, to the neglect of intercommunity relationships on a micro-scale.

Recent studies on the Lesser Antilles have provided evidence for the existence of such relationships through the distribution and circulation of raw materials and exotic items. Some of the most widespread and useful studies to help determine the provenance and distribution of goods between various islands are the analyses of lithics and clays used to manufacture pottery (for example, Carini 1991; Descantes et al. 2005; Hofman et al. 2006c; Knippenberg 2006; Serrand 1999; Vescelius and Robinson 1979; Watters and Scaglion 1994). These studies suggest that local groups were interacting on a fairly frequent basis, although the extent to which they exchanged goods (but also knowledge or marriage partners, for example) is not well understood. As such, these studies have not considered a framework of interaction and exchange in which the dynamics of the material and social and ideological relationships are interwoven.

The present study is part of a multiyear program on analyzing mobility and exchange in the circum-Caribbean region and more particularly in the Lesser Antilles. In this study, the provenance and distribution patterns of lithic raw materials, pottery, and exotic items are investigated in relation to the mobility of people and the subsequent transfer of knowledge and ideas. An interdisciplinary approach using archaeological, archaeometric, and ethnohistorical data should provide an integrated view of the interaction networks evinced by the pre-Columbian inhabitants of the circum-Caribbean (Hofman and Hoogland 2004; Hofman et al. 2006b).

In this chapter, we focus on the mobility of prehistoric peoples on the micro-scale of the Lesser Antilles based on the identification of local and nonlocal[1] individuals in a large and diverse burial assemblage from the site of Anse à la Gourde, Guadeloupe. The site was excavated between 1995 and 2000 as a cooperative effort between Leiden University and the Direction Régionale des Affaires Culturelles de la Guadeloupe. The excavations revealed over eighty burials with a total of ninety-three individuals, one of the largest cemeteries yet discovered in the Lesser Antilles. Strontium isotope analysis on fifty of the individuals provides evidence that a significant portion of the population had a nonlocal origin. This would suggest that intercommunity mobility was a frequent, and probably necessary, component to island lifeways during the Late Ceramic Age, that is, between AD 800 and 1492 (see also Booden et al. 2008).

## Studies on Mobility in the pre-Columbian Caribbean

The maritime orientation of Lesser Antillean communities was extremely conducive to maintaining extensive and intensive contacts across the entire region. The sea, circumscribing islands on all sides, helped link island communities to each other, encouraging mobility and exchange (Hofman et al. 2006c; Watters 1997; Watters and Rouse 1989). On a macro-scale, extensive population movements through the islands have been documented by cranial and dental morphological traits and corroborated through the analysis of ancient DNA of various Archaic and Ceramic Age burial assemblages in the Greater and Lesser Antilles (chapters 3, 5, and 9, present volume; Coppa et al. 2003, 2006; Lalueza-Fox et al. 2003). Strontium (Sr) isotope analysis has, however, hardly been applied in the circum-Caribbean.

The application of strontium isotope analysis on human dental elements from Europe (Evans et al. 2006; Schweissing and Grupe 2003), the southwestern United States (Ezzo and Price 2002; Ezzo et al. 1997; Price et al. 1994), and Mesoamerica (Hodell et al. 2004; Price et al. 2000; Wright 2005) has shown the value of this method to investigate cases of migration and population movement throughout these regions. In the same vein, a study of $^{87}Sr/^{86}Sr$ ratios in tooth enamel distinguished between local members of a community and immigrants in Peru (Knudson et al. 2004) and Bolivia (Knudson et al. 2005) on the basis of regional geological variation.

The geological makeup of the islands of the Lesser Antilles, comprising an inner arc of volcanic islands and an outer arc of limestone islands, offers good opportunities for applying strontium isotope analysis to the study of intraisland human mobility. During the Late Ceramic Age, microregional interisland networks are evidenced between these islands on the basis of the circulation of goods. It is assumed that the exchange of raw materials and finished goods is concomitant with human mobility. Descent and marital residence practices could be the underlying mechanisms for such exchange relationships.

In general it has been assumed that matrilineal communities with a matrilocal residence rule were predominant in Antillean societies. Keegan (2009) has recently proposed that residence patterns may have shifted from matrilocal toward avunculocal at the end of the Saladoid (circa 500 BC–AD 600). Based on ethnohistorical accounts of the Taíno, Keegan and Maclachan (1989) suggested that viri-avunculocality was the principal residence rule among elites. However, Curet (2002) argued that these conclusions were based on uncertain assumptions and that rules of succession were being confused with rules of descent. Keegan has since shown that succession was based on matrilineal descent and that this form of descent reckoning must have been common in the Taíno societies of Hispaniola (Keegan 2006b).

Descent and residence rules have major implications for mortuary practices (that is, the composition of the burial assemblage and the spatial distribution of the burials in a cemetery). In a recent article, Keegan (2009) proposed that we consider the possibility for postmortem mobility (that is, people born in one village, living their adult life in another village, and returning to their home or clan village for burial). The picture emerging from these studies underlines the dynamic nature of insular communities adapting and manipulating descent and residence rules for sociopolitical purposes. Archaeometric studies such as isotope analysis and extraction of ancient and modern DNA are indispensable to disentangle the Gordian knot of pre-Columbian Antillean descent and residence rules.

## Archaeological Background

### Anse à la Gourde

Anse à la Gourde is situated in the northeastern part of Grande-Terre in the northern Lesser Antilles. The site is located on a limestone plateau and has a surface area of approximately four hectares (ha). Occupation of the site took place from cal AD 450 to 1420, based on a suite of radiocarbon dates. The site was most densely occupied between cal AD 1000 and 1250 and consisted of a habitation area surrounded by a doughnut-shaped re-fuse midden. A total of 2,400 features belong to both houses and other do-mestic structures, hearths, refuse pits, and burials. Most burials are located within an area with a high concentration of postholes.

Ceramics belong to the late Cedrosan Saladoid, Mamoran and/or Trou-massan Troumassoid, and Suazan Troumassoid subseries (Hofman et al. 1999, 2001; see also Petersen et al. 2004). The heterogeneity of ceramic styles suggests that there were both local and regional contacts. Prov-enance studies have confirmed that mostly local sources were used for the production of lithic, shell, and coral artifacts and pottery. However, there is clear evidence that some artifacts have a nonlocal provenance. As such, there are axes and adzes made of greenstone from St. Martin, three-pointers manufactured of calci-rudite from that same island, tools of flint from Antigua, and polishing pebbles of a greenstone from La Désirade (Knippenberg 2006). Technological and functional studies of the shell or-naments also provided evidence for the nonlocal manufacture of some of the beads (Lammers-Keijsers 2007).

### Burial Assemblage

A total of eighty-three human burials contain the remains of ninety-three individuals. For the determinations of sex and age, the methods stipu-lated by the Workshop of European Anthropologists (WEA 1980) have been applied, as published in the Manual for the Physical Anthropological Report (Maat et al. 2002). The assemblage is composed of mostly adult individuals. Of the individuals whose sex could be determined, 38 percent are males and 62 percent are females, suggesting high female mortality. An alternative explanation for the underrepresentation of males in the burial assemblage could be their death abroad during trading and raiding expeditions (Ember and Ember 1971). Contrary to what one would expect from the demographic models, there is a low number of juveniles (11 percent) buried among the adults within the habitation area. Children could well have been considered to belong to a different social category and therefore received a different mortuary treatment than adults.

Mortuary Practices

At Anse à la Gourde, the various and complex modes of inhumation include primary and secondary burials as well as both single and composite burials (Hoogland et al. 1999). Males tend to be buried semiseated or seated, suggesting that they may have been buried sitting on a wooden stool, and females in a dorsal position with flexed lower limbs.

Mortuary practices include the removal, intentional fracturing, and secondary deposition of bones, as well as the interment of skeletal remains of an additional individual in a grave and the displacement of bones within the context of the grave. In some 25 percent of the burials, crania were removed, leaving the mandible or only dental elements in the grave. These practices also occur in various combinations. In a number of cases, the body of the deceased seems to have been prepared before interment by being wrapped in a bundle (using a hammock), for example, and then allowed to dry. Regardless of how the body was prepared for burial, it was evidently always deposited in a shallow pit. The ash spots in and around some of the graves at both sites indicate that a fire was burned near or on the grave. Historical sources suggest that such fires had a ceremonial function (Conseil Général de la Martinique 2002 [1694]: 125) or served to incinerate the perishable personal belongings (Breton 1665: 80–81). An ethnographic account by Roth (1924: 659–64) of the Makusi in Guyana relates that after the deceased was buried in a open grave, all relatives left the house, except for an old female relative who took care of the small fire burning next to the grave for weeks.

There are some burials with postholes around the grave pit, as well as burials with vessels covering a large part of the body. The pottery vessels associated with the graves are either 50–60-centimeter-wide plates used as a lid to cover the open grave pit or bowls placed in front of the head of the deceased covering the face. Some artifacts, mostly shell beads, seem to

Table 7.1. Radiocarbon Dates of Nine Individuals of the Burial Assemblage of Anse à la Gourde

| Lab code | Description | Radiocarbon age | σ ranges | Relative area under probability distribution | 2σ ranges | Relative area under probability distribution |
|---|---|---|---|---|---|---|
| GrN-22795 | AAG F108 | 1030±40 | cal AD 1049–1152 | | cal AD 1024–1192<br>cal AD 1198–1205 | .985861<br>.014139 |
| GrN-22796 | AAG F311 | 1000±25 | cal AD 1053–1089<br>cal AD 1120–1183 | .342533<br>.657467 | cal AD 1048–1209 | |
| GrN-22797 | AAG F350 | 950±50 | cal AD 1153–1261 | | cal AD 1050–1108<br>cal AD 1116–1271 | .148516<br>.851484 |
| GrN-22798 | AAG F378 | 910±25 | cal AD 1217–1261 | | cal AD 1181–1274 | |
| GrN-26151 | AAG F2217 | 770±45 | cal AD 1283–1322<br>cal AD 1356–1388 | 0.583768<br>0.416232 | cal AD 1272–1401 | |
| GrN-26152 | AAG F2216 | 720±40 | cal AD 1310–1371<br>cal AD 1379–1395 | 0.801601<br>0.198399 | cal AD 1293–1412 | |
| GrN-26153 | AAG F2212 | 910±40 | cal AD 1206–1270 | | cal AD 1159–1282 | |
| GrN-26154 | AAG F2213 | 880±40 | cal AD 1224–1275 | | cal AD 1179–1292 | |
| GrN-26155 | AAG F1126B | 760±40 | cal AD 1288–1323<br>cal AD 1355–1388 | 0.518458<br>0.481542 | cal AD 1279–1399 | |

*Note:* Calibrated by CALIB 5.0.2 (Stuiver and Reimer 1993) using a "mixed" marine and northern atmospheric calibration curve. It was assumed that the percentage of marine food in the diet of this population is a minimum of 30 percent. A Delta-R of -29±20 years was used.

have been part of a garment or ornament. In a few cases, grave goods could be interpreted as personal property, such as a set of twenty-one *Strombus* pebbles (Lammers-Keijsers 2007: 72–73), a flint core, and two greenstone celts.

Within the habitation area, burials occur in clusters, with each cluster comprising three to ten burials. The close association of the postholes with the burials provides an indication that the majority of the burials are located under the house floors. The latter points to a shift from the foregoing Early Ceramic Age (Saladoid) burial assemblages in the Lesser Antilles, in which the deceased were buried outside the houses, to burials within residential structures (see, for example, Versteeg and Schinkel 1992). This shift, also noted for other parts of the Caribbean, has been interpreted as related to changes in sociopolitical structure and to an increase in ancestor veneration during the Late Ceramic Age (Curet and Oliver 1998; Hofman et al. 2001; Siegel 1996). Radiocarbon dates for the burials at Anse à la Gourde range between cal AD 1020 and 1410 (see table 7.1).

## Strontium Isotope Analysis of Burials at Anse à la Gourde

In most recent strontium isotope studies, the human sample material is restricted to tooth enamel samples since bone is often contaminated by strontium from the soil (Bentley 2006). In the case of Anse à la Gourde, only tooth enamel samples were included in the study because of the moderate preservation of the skeletal remains (Booden et al. 2008). One tooth, preferably a premolar to ensure a comparable age of formation for each sample, was selected from each individual. Because the enamel samples were taken from premolars, which are formed during childhood, at the age of approximately six years, the interpretations of the timing of the mobility would ultimately be limited to sometime after six years of age.

The selected sample consisted of fifty teeth that were free of visible cracks, lustrous, and either white or covered by only a thin and easily removed outer layer of dull, yellowed enamel with white enamel underneath. Strontium isotope ratios were determined using a FinniganMat 262 RPQ Plus thermal ionization mass spectrometer at the Vrije Universiteit (Free University), Amsterdam, The Netherlands. The methodological aspects of the analysis of strontium isotope ratios ($^{87}Sr/^{86}Sr$) in enamel have previously been reported by Booden and colleagues (2008). The $^{87}Sr/^{86}Sr$ values range from 0.707490 in the individual of burial F378 to 0.709412 in the individual from burial F2211 (table 7.2).

Table 7.2. Strontium Ratios for Human Teeth ($N$=50) and Rice Rat Molars ($N$=4) (after Booden et al. 2008)

| Feature | Label | $^{87}Sr/^{86}Sr$ | 2 SD |
|---|---|---|---|
| F0050 | Local human | .709171 | .000008 |
| F0089 | Local human | .709161 | .000007 |
| F0108 | Local human | .709172 | .000015 |
| F0139 | Local human | .709034 | .000009 |
| F0159 | Local human | .709146 | .000009 |
| F0171 | Nonlocal human | .708636 | .000020 |
| F0196 | Local human | .709038 | .000009 |
| F0202 | Local human | .709127 | .000012 |
| F0206 | Local human | .709127 | .000009 |
| F0207 | Local human | .709116 | .000013 |
| F0212 | Local human | .709083 | .000010 |
| F0241 | Local human | .709058 | .000008 |
| F0253 | Local human | .709162 | .000015 |
| F0288 | Nonlocal human | .708646 | .000008 |
| F0292 | Nonlocal human | .708755 | .000010 |
| F0304 | Local human | .709122 | .000040 |
| F0307 | Local human | .709165 | .000008 |
| F0311 | Nonlocal human | .708849 | .000014 |
| F0332 | Nonlocal human | .708278 | .000009 |
| F0342 | Local human | .709031 | .000007 |
| F0349C | Nonlocal human | .708590 | .000009 |
| F0350 | Local human | .709182 | .000010 |
| F0377 | Local human | .709071 | .000027 |
| F0378 | Nonlocal human | .707490 | .000009 |
| F0430 | Nonlocal human | .708794 | .000009 |
| F0447 | Local human | .709090 | .000011 |
| F0450 | Nonlocal human | .708690 | .000009 |
| F0451 | Local human | .709113 | .000008 |
| F0452 | Local human | .709182 | .000006 |
| F0454 | Local human | .709164 | .000009 |
| F0706 | Local human | .709168 | .000018 |
| F0726 | Local human | .709228 | .000013 |
| F0953 | Local human | .709162 | .000012 |
| F1204 | Local human | .709131 | .000012 |
| F1226 | Local human | .709068 | .000009 |
| F1413 | Local human | .709193 | .000018 |
| F1496 | Local human | .709158 | .000007 |
| F1651 | Local human | .709168 | .000006 |
| F1944 | Nonlocal human | .709287 | .000010 |
| F1945 | Local human | .709166 | .000007 |
| F1947 | Local human | .709237 | .000013 |
| F1948 | Nonlocal human | .708895 | .000009 |
| F2005 | Nonlocal human | .708475 | .000012 |
| F2107 | Local human | .709149 | .000008 |
| F2211 | Nonlocal human | .709412 | .000014 |
| F2212 | Local human | .709100 | .000008 |
| F2214 | Local human | .709156 | .000012 |
| F2215 | Nonlocal human | .707747 | .000007 |
| F2216 | Local human | .709165 | .000006 |
| F2217 | Local human | .709168 | .000005 |
| k617 | Rat | .709205 | .000011 |
| k619 | Rat | .709207 | .000007 |
| k623 | Rat | .709132 | .000009 |
| k628 | Rat | .709149 | .000013 |

Table 7.3. Mean Value of Strontium Ratios for Rice Rat Molars, Human Teeth, and Soil Samples (after Booden et al. 2008)

|  | Number of samples | Mean $^{87}Sr/^{86}Sr$ | SD |
|---|---|---|---|
| Rice rats | 4 | .709173 | .000038 |
| Soil | 4 | .709156 | .000030 |
| All individuals | 50 | .708989 | .000359 |
| Local individuals | 36 | .709137 | .000051 |

In other studies, $^{87}Sr/^{86}Sr$ ratios in the teeth of small or domesticated animals have been used as well as ratios in soil, bedrock, and surface or ground water to determine local strontium isotope signatures (Bentley et al. 2004; Hodell et al. 2004; Price et al. 2002). Skeletal remains of rice rats (*Oryzomys* sp.) are rather common in Ceramic Age midden deposits of archaeological sites in the Caribbean. It was expected that rice rats would be a good indicator for the site-specific $^{87}Sr/^{86}Sr$ ratios in humans, as the rice rats would have fed in the neighborhood of settlements on the refuse deposits and in the nearby (kitchen) gardens. Incisors from four mandibles of rice rats and four samples of soil collected from the graves provide the local $^{87}Sr/^{86}Sr$ baseline of 0.709165 (Booden et al. 2008). The sample was subjected to statistical analysis. A subset of individuals showing similar signatures was distinguished from the individuals with divergent signatures by iteratively excluding outliers from the data set, until no further outliers are identified. Outliers were defined as values outside the 95 percent confidence interval (Booden et al. 2008). The outliers were interpreted as the nonlocal individuals in the burial assemblage of Anse à la Gourde, and the individuals with $0.70903 > {}^{87}Sr/^{86}Sr > 0.70924$ are defined as locals. The local population has a mean of 0.709137, falling well within the range of enamel from rice rats and soil samples (table 7.3).

Results indicate that in 28 percent ($n = 14$) of the individuals, the strontium signature of the dental elements shows a nonlocal provenance (figure 7.1). As defined by Bentley (2006; Bentley et al. 2004) a "nonlocal" is an individual whose strontium isotope signature of the dental enamel falls outside of the local range of strontium isotope variation for a given region. In this case of Anse à la Gourde, the direct surroundings of the site are geologically characterized by a bedrock of young limestone, which possesses a fairly homogenous strontium isotope signature similar to that of modern seawater (~.7092). Similar limestone sediments occur on two islands of the Guadeloupean archipelago, that is, Grande-Terre and Marie-Galante,

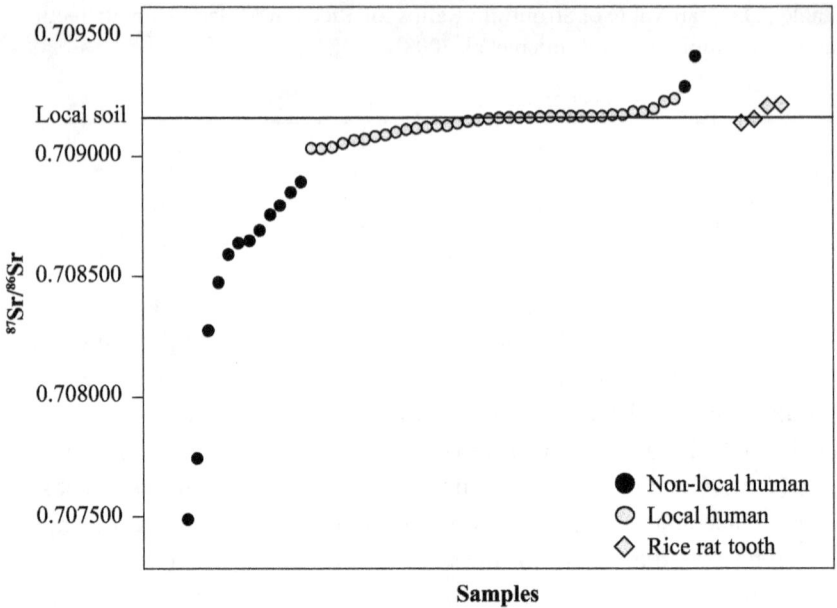

Figure 7.1. Strontium isotope ratios of the individuals at Anse à la Gourde (after Booden et al. 2006).

as well as on Barbuda and Anguilla in the northern Lesser Antilles. This means that individuals with isotopic ranges substantially different from .709125±.00001 were not raised in any region with young limestone environments, including Grande Terre, Marie-Galante, or any other region with similar deposits.

All fourteen nonlocal individuals at Anse à la Gourde appear to originate from areas of nonlimestone geology, probably from an island with volcanic deposits or an island with composite volcanic-limestone structure, such as Antigua, St. Barthelemy, and St. Martin. The variation in the strontium isotope ratios of the nonlocals indicates that they originate from at least three different geological settings. Determination of local faunal signatures for the Lesser Antillean islands is ongoing and may help to further narrow down the potential geographical origins of these nonlocal individuals. There are, however, a number of constraints in tracing the origins of humans, with regard to the marine component in their diet. The latter raises the strontium isotope ratio and together with the rising effect of sea spray can have an important impact on local signatures in coastal settlements on volcanic and composite islands.

## Grave Inventory of Nonlocal Individuals

The nonlocal individuals are randomly distributed over the habitation area. They consist of four adult males, seven females, and three adults of undetermined sex. Burial practices of these nonlocal individuals include primary and secondary burials, both simple and composite. Of the fourteen nonlocal individuals, eight (one male, five females, and two of undetermined sex) have objects deposited in the grave, including shell artifacts (like *Chama sarda* beads), coral artifacts, a perforated shark tooth, lithic artifacts, and a pottery vessel in front of the face. Two of the three lithic artifacts are manufactured of St. Martin greenstone, and one is made of Antigua flint. A bead belt or apron made from more than one thousand *Strombus* beads is also of nonlocal fabrication (Lammers-Keijsers 2007). The beads were located on the pelvis of an adult female (20–25 years old) (figures 7.2a–b, 7.3). Such artifacts have not been found with local individuals.

## Conclusions

At Anse à la Gourde, microregional mobility of a Late Ceramic Age population is suggested on the basis of the results of strontium isotope analysis. Of the fifty individuals included in the study, 28 percent showed nonlocal signatures. The majority of them were females. They were born elsewhere and then moved to Anse à la Gourde, sometime after six years of age, where they were also buried.

On the basis of the spatial distribution of the burials of the nonlocal individuals in the site and their strontium isotope heterogeneity, it is unlikely that they represent one single group of migrants. They appeared to have been buried randomly in the habitation area amidst local individuals. Moreover, there seems to be no significant difference in mortuary practices between the local and nonlocal individuals. Analysis of grave inventories, however, shows that grave goods of the nonlocal individuals were manufactured from nonlocal materials. All of the nonlocal objects were exclusively found with nonlocal females or with individuals of undetermined sex. This unique occurrence of nonlocal female individuals buried with nonlocal grave goods may offer a rare insight into direct transmission, that is, the carrying of material culture directly by the people concerned, as opposed to the exchange of the goods from hand to hand.

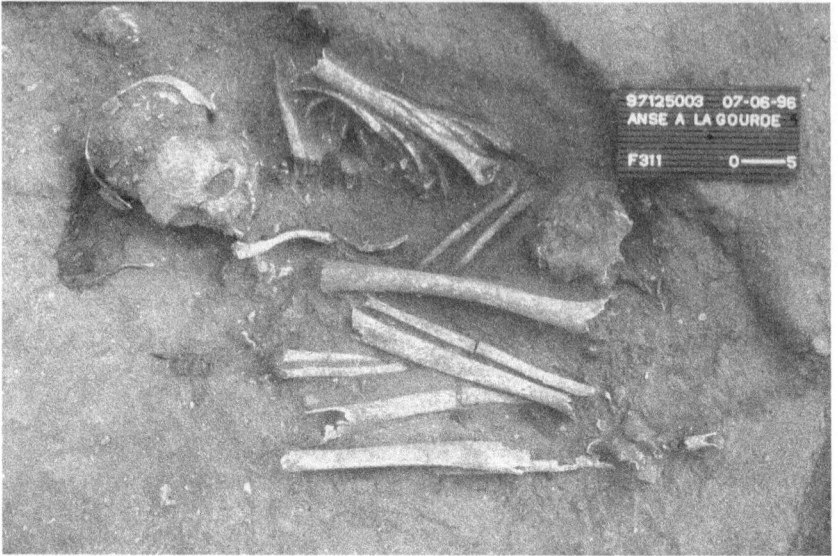

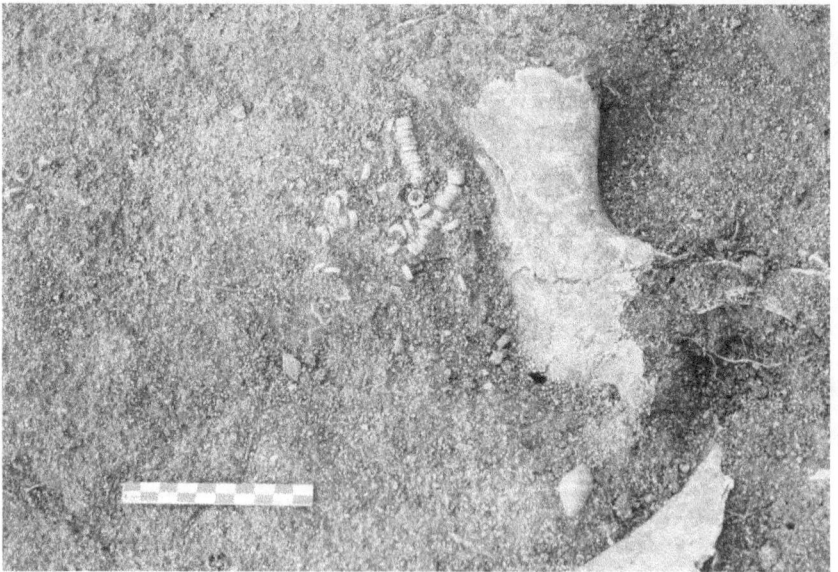

Figure 7.2a and b. Nonlocal female burial with more than one thousand *Strombus* beads of nonlocal fabrication on her pelvis (photograph by Menno Hoogland).

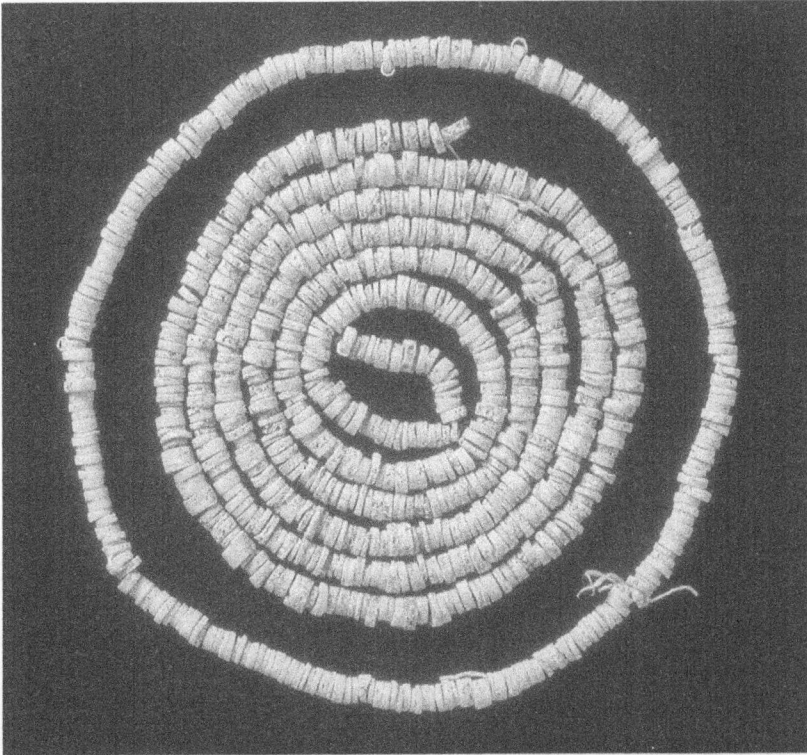

Figure 7.3. *Strombus* beads of nonlocal fabrication (photograph by Jan Pauptit, courtesy of Corinne Hofman and Menno Hoogland).

Females also represent the majority among the nonlocal individuals. This trend may be indicative of a preference for virilocal above matrilocal marital residence and as such would confirm Keegan's assumption that in Late Ceramic Age communities there is a trend of development from matrilocal marriage residence toward virilocal residence among members of the local elite while maintaining the principle of matrilinear descent. The necessity to concentrate on male relatives in one domestic unit could eventually lead to avunculocal marital residence. In this situation, the married matrilineage males remain home by joining the residence of their mother's brother, implying a relatively high degree of female mobility. In societies with a high male mortality due to warfare, there tends to be a shift from matri- to virilocality. Although as yet there are no documented indications for intensive warfare in the Anse à la Gourde burial assemblage, this remains an option to be explored in the future.

For the thirty-six identified local individuals, three possibilities may be suggested. First, all individuals were born at Anse à la Gourde, spent their entire life living at or near the site, and were also buried there. A second option is that some of the individuals originated from a place with strontium isotope signatures similar to those of Anse à la Gourde, such as other coastal settlements in the northeastern part of Grande-Terre or other neighboring limestone islands. A third possibility is that some of the individuals with local signatures were born at Anse à la Gourde, spent their adult lives elsewhere, and were returned to their ancestral village after death. This would entail postmortem mobility (Rubel and Rosman n.d.; Keegan pers. comm. 2006), an issue that could only be resolved by obtaining strontium isotope signatures from human bone. These signatures offer an insight into a person's adult life residence, as bone is regenerated on average every ten years, meaning its strontium isotope ratios can change during lifetime.

The refinement of the method (more specifically, finding a solution for the problem of diagenesis and the further development of a database for individual and local signatures) is essential to further elucidate all aspects of human mobility. Moreover, it offers the possibility to disentangle the mechanisms underlying the sociopolitical structure and the complicated web of social relationships that tied together the early insular inhabitants of the Lesser Antilles and the circum-Caribbean as a whole.

## Acknowledgments

The strontium analysis has been carried out in the scope of the VIDI project "Mobility and Exchange," financed by the Netherlands Foundation for Scientific Research (NWO nr. 276.62.001). The samples were processed by Mathijs Booden, Marin Waaijer, and Hayley Mickleburgh in the laboratory facilities at the VU at Amsterdam under the supervision of Prof. Dr. Gareth Davies. Jason Laffoon and Hayley Mickleburgh are acknowledged for their enjoyable discussions on the subject and stylistic advice. Finally, the authors are indebted to William Keegan, Scott Fitzpatrick, and an anonymous reviewer for their useful remarks and suggestions that certainly improved the text.

## Notes

1. The term *nonlocal* is preferred above others, leaving open a wide array of mechanisms underlying mobility, such as migration, exchange of marriage partners, captivity of slaves, war trophies, or (re-)burial of individuals in their ancestral burial ground.

# 8

# The Southward Route Hypothesis

## Examining Carriacou's Chronological Position in Antillean Prehistory

SCOTT M. FITZPATRICK, MICHIEL KAPPERS,
AND CHRISTINA M. GIOVAS

## Abstract

The prevailing hypothesis for explaining the migration of what appear to be the first ceramic-making people (Saladoid) into the Caribbean has been that they moved northward from South America into the Lesser Antilles in a stepping-stone fashion around 500/400 BC. This seems to be corroborated by a corpus of relatively contemporaneous $^{14}$C dates and similarities in ceramic styles from Puerto Rico east and south through the Lesser Antilles. However, recent studies using computer simulations of seafaring and a reevaluation of radiocarbon chronologies suggest that peoples may have first voyaged directly to the northern islands and then moved southward. To examine this hypothesis and to help construct a more detailed chronology for islands in the southern Lesser Antilles, we present a suite of twenty new radiocarbon dates from Carriacou in the Grenadines. Results of $^{14}$C dating of the two largest archaeological sites on Carriacou, Grand Bay and Sabazan, indicate that the earliest occupation of the island was cal AD 400–600, suggesting that peoples may have initially reached islands farther north and then moved south down through the Lesser Antilles over the next few hundred years.

## Resumen

La hipótesis predominante para explicar la migración de las primeras culturas ceramistas (Saladoide) en el Caribe ha sido que ellos se desplazaron hacia el norte de Sudamérica en las Antillas Menores en forma de pasadera alrededor de 500/400 AC. Esto parece ser corroborado por fechas relativamente contemporáneas $^{14}$C y similitudes en estilos de cerámicas del este de Puerto Rico y al sur por las Antillas Menores. Sin embargo, los estudios recientes que utilizan simulaciones por ordenador de navegación y una reevaluación de cronologías de radiocarbono sugieren que personas pueden haber viajado directamente a las islas septentrionales y después progresado hacia el sur. Para examinar esta hipótesis y para ayudar a construir una cronología más detallada para islas en las Antillas Menores meridionales, nosotros presentamos una serie de veinte nuevas fechas de radiocarbono de Carriacou en las Granadinas. Los resultados de $^{14}$C que fecha de los dos sitios arqueológicos más grandes en Carriacou, la Bahía Grandes y Sabazan, indican que la ocupación más temprana de la isla fue cal. d.C. 400–600, sugiriendo que personas pueden haber alcanzado inicialmente las islas de extremo norte y después se desplazaron hacia el sur por las Antillas Menores.

## Introduction

One of the primary goals of archaeology in the Caribbean has been to trace prehistoric patterns of human migration and settlement (Rouse 1989). Research demonstrates that at least three major migrations occurred in the region. Among the earliest of these, termed the Lithic or Casimiran Casimiroid (Rouse 1992a) and found only in Cuba and Hispaniola, dates to as early as circa 3000 BC (Pino and Castellanos 1985; see also Jull et al. 2004: 670; Steadman et al. 2005: 11766) and may have originated from somewhere in Mesoamerica (Keegan 1994, 2000; Wilson et al. 1998). Earlier Archaic Age, or "preceramic," peoples occupied Trinidad (possibly circa 5000 BC) and islands adjacent to the South American mainland (for example, Aruba and Curaçao) circa 1300 BC (Haviser 1991), when sea levels were lower (Boomert 2000). About 1,500 years later, these Archaic (or Ortoiroid; see Rouse 1992a) peoples entered the Lesser Antilles from South America (Keegan 1994; Callaghan 2003c) as evidenced from numerous islands, including Antigua (Davis 2000) and Saba (Hofman et al. 2006a).

This was later followed, around 500/400 BC, by ceramic-making horticul-turalists known as Saladoid, whose distinctive pottery persists until circa AD 600 and stretches from Trinidad to Puerto Rico, linking the ancestry of these peoples in some fashion to northern South America (Rouse 1976, 1986, 1989; Keegan 1994).

In attempting to describe the structure and behavior of prehistoric mi-grations during the Ceramic Age, archaeologists have often subscribed to the notion of a "stepping-stone" movement whereby peoples are thought to have traveled northward from South America up through the Antillean chain of islands, eventually reaching Puerto Rico around 500/400 BC (see, for example, Rouse 1989, 1992a: 79; Drewett 2000). On the surface this ap-pears valid from a biogeographical perspective—not only are many of the islands in the Antilles close together, but in clear weather they are visible from one to the next or can be seen from a point midway between two islands. In addition, early Saladoid pottery has been reported from Puerto Rico and islands throughout the Lesser Antilles, suggesting that the east-ernmost islands in the Caribbean were settled by the same related group relatively quickly.

A northward stepping-stone migration has been widely accepted but remains poorly tested. The idea is now waning in popularity due to a more concerted effort by archaeologists to critically examine radiocarbon chronologies (chapter 2, this volume; Keegan 2000; Fitzpatrick 2006) and conduct computer simulations of voyaging (Callaghan 2001, 2003c). In general, however, there has been no real attempt to examine either hy-pothesis, which requires the submission of multiple radiocarbon samples from stratified deposits within well-documented archaeological sites and a critical assessment of whether the dates are acceptable (Fitzpatrick 2006). Although sites on one island may appear to be the same age as those on others, these observations are based not on carefully constructed radio-carbon chronologies but on the artifacts themselves, similar ones of which have been dated, sometimes loosely, on other islands (for example, Haviser 1997; see Keegan 1994: 263). This overlooks the possibility that unique sty-listic attributes of pottery can persist for several hundred years, that they can begin and end at different times on different islands, and that peoples could have bypassed some islands in the Antillean chain in whatever di-rection they may have taken (see Boomert 2000: 222–24; Hofman 1993). The idea of a stepping-stone migration has been so pervasive, in fact, that many archaeologists have had a tendency to simply accept this hypothesis and fit their data into this pattern automatically without submitting multiple (or any) [14]C dates from individual strata or sites to associate with archaeologically recovered material (Fitzpatrick 2006). As a result, evidence to the contrary has not been fully explored and alternative hypotheses remain underdeveloped.

One of these alternative hypotheses, suggested by a growing corpus of data, is that Saladoid colonizers jumped directly from the northern coast of South America to Puerto Rico and the northern Lesser Antilles and then worked their way southward through the Leeward and Windward islands, bypassing some islands in between. This is in keeping with the distribution of early Saladoid sites in the Caribbean, which are concentrated in the north (for example, Puerto Rico, Vieques, St. Martin, Montserrat) (see Haviser 1997: table 7.1, figs. 7.1 and 7.2), and with voyaging simulations, which indicate that the patterns of the currents and winds in the West Indies favored this route (Callaghan 2001, 2003c). This southward route is emerging as a viable alternative hypothesis to the stepping-stone model of migration but, as yet, remains largely unassessed.

In this chapter, we describe the effort to address this problem on Carriacou, a small but archaeologically rich island in the Grenadines chain of the southern Caribbean. Archaeological investigations on the island since 2003 have revealed large and well-stratified anthropogenic deposits at several sites that make them extremely suitable for examining colonization strategies and settlement patterns on the island and in the region. Radiocarbon dates, primarily from the sites of Grand Bay and Sabazan, have provided a good sequence of later Ceramic Age habitation but no chronological evidence for what would be the first nine hundred years of Saladoid occupation (circa 500/400 BC–AD 400). Although Saladoid-era pottery such as ZIC (zone incised crosshatch) ware has been identified on Carriacou (Harris 2005), these dates, along with evidence from other sites in the region, seem to support the hypothesis that ceramic-making peoples probably reached the northern Antilles first and then migrated southward, while still continuing the long-held tradition of making pottery with characteristics common to the Saladoid period.

We first briefly describe archaeological work conducted on Carriacou and then provide a chronology for the island based on a suite of twenty-one (sixteen new, four previously reported in Fitzpatrick et al. 2004, and one reported in Bullen and Bullen 1972) radiocarbon dates from the sites of Grand Bay, Sabazan, and Harvey Vale. We then contextualize these findings into a broader perspective for the region's prehistory, arguing that the current data, in conjunction with other $^{14}$C dates and artifactual evidence, currently do not support a south-to-north stepping-stone migration of Saladoid peoples into the Lesser Antilles.

## Archaeological Background

Carriacou is the largest island in the Grenadines chain in the southeastern Caribbean. The island is located approximately 200 kilometers north of Venezuela and 30 kilometers north of Grenada; it measures 10 kilometers from north to south and 8 kilometers across at its widest point and is roughly 32 square kilometers in area (figure 8.1). Geologically, Carriacou is composed of a mixture of volcanic lava and Miocene-aged fossiliferous limestone (Pickerill et al. 2001, 2002; Donovan et al. 2003) that reach

Figure 8.1. Map of Carriacou with identified archaeological sites.

heights of up to 290 meters above sea level in both the island's northern and southern half.

One of the first attempts to investigate Carriacou was by Jesse Fewkes (1907: 189–90) in 1904; he described the artifacts found there as "among the finest West Indian ware that has yet come to the Smithsonian Institution." Intensive archaeological research on the island, however, has been limited. Bullen and Bullen (1972) made a day-long trip to Carriacou in 1964 to collect artifacts and excavate a foot-thick "slice" from the coastal profile at Sabazan. Sutty (1990) conducted a more detailed survey of Carriacou and recorded a number of sites with a wide array of ceramic styles, some of which appeared to have unique designs.

To begin investigating the pre-Columbian settlement of Carriacou, we have worked as part of an international team of British, Dutch, and American archaeologists to systematically record sites on the island (Kaye 2003). Eleven locations with evidence for occupation were recorded, six of which have finds indicative of long-term settlement (Kaye et al. 2005; figure 8.1). Of these six sites, Sabazan and Grand Bay have the most extensive stratified coastal profiles and an abundance of faunal remains, artifacts, postholes, hearth features, and human burials. In addition, both sites were, and still are, heavily eroding into the sea, so it was critical that we begin efforts to record these sites before they eroded away completely (Fitzpatrick et al. 2006).

Grand Bay and Sabazan are both located on the southeastern part of the island and consist primarily of dense middens. Grand Bay is the largest site, covering approximately 6,000 square meters in area, and is intercut by a series of eroded gulleys (Fitzpatrick et al. 2006, 2010) with an overlying humic topsoil layer and yellowish brown subsoil into which a number of burial and household features intrude. Sabazan's coastline stretches for about 100 meters and has a similar series of midden deposits overlaying subsoil. The exact boundaries of Sabazan have not yet been determined, although it is at least somewhere between 2,500 and 3,500 square meters, based on initial survey results. To help record these sites and determine the earliest occupation of Carriacou and subsequent periods of activity, we submitted samples from these two sites as well as Harvey Vale (Tyrell Bay) for radiocarbon dating.

## Radiocarbon Dating

A total of twenty samples from Grand Bay ($n = 8$) and Sabazan ($n = 12$) have been radiocarbon dated to complement the single date (RL-29) reported by Bullen and Bullen (1972). An additional sample from a human skeleton recovered by members of the Carriacou Historical Society (CHS) Museum in Harvey Vale was also submitted to determine its antiquity. All samples from Grand Bay and Sabazan were collected with a clean trowel, then individually bagged, dried, and placed in waterproof containers. The samples were washed thoroughly with distilled water, dried, double-sealed in bags and film canisters, and submitted to four different laboratories for both accelerator mass spectrometry (AMS) and conventional radiocarbon dating.

A charcoal sample (OS-41358) from Sabazan was sent to the National Oceanic Sciences AMS (NOSAMS) laboratory in Woods Hole, Massachusetts; eight charcoal samples were submitted to the University of Arizona's AMS lab; and three marine shell samples were sent to Geochron Laboratories, Inc., in Cambridge, Massachusetts, for conventional dating (GX-30423, 30424, and 30425) (table 8.1). Samples from Sabazan were collected from various cultural layers along the tallest sections of the profile (see Fitzpatrick et al. 2004: 5, fig. 3). The middle to upper strata (generally the top 2 to 2.5 meters) had extensive cultural remains (for example, pottery, vertebrate bone, invertebrates, and charcoal). The main occupational strata were clustered together from roughly 1.5 to 2.5 meters in depth and were primarily sandy clays with rich humic topsoils intermixed with inclusions of coral rock and other debris (see Fitzpatrick et al. 2004).

The samples from Grand Bay and Harvey Vale were sent to the University of Arizona's AMS Facility and Beta Analytic, Inc. (Miami, FL). The Grand Bay samples included marine shell, charcoal, and human bone. Four of the samples (two charcoal and two marine shell) were collected from the southern coastal profile (figure 8.2) at depths ranging from 93 to 145 centimeters across two successive stratigraphic layers. One of the shell samples (AA-62280) was dated twice by the lab as an internal measure to ensure that the samples were undergoing satisfactory pretreatment and analysis within acceptable levels. Based on our observations of surface scatter and eroding ceramic material at the site, it appeared that the earliest occupation may have occurred along the northern profile. Therefore, we submitted a juvenile queen conch (*Strombus gigas*) shell taken from the basal floor of the lowest occupied strata to compare with that of the

Table 8.1. Radiocarbon Dates from Carriacou

| Sample no. | Site | Lab no. | Type | Species | Unit | Layer | Cmbs | $^{13}C/^{12}C$ ratio | Measured $^{14}C$ age BP | Cal. BC/AD (2σ) |
|---|---|---|---|---|---|---|---|---|---|---|
| CAR-1 | Sabazan | GX-30423 | Shell | C. pica | Profile | VI | 160 | 2.4 | 1,400±60 | AD 870–1160 |
| CAR-2 | Sabazan | GX-30424 | Shell | S. gigas | Profile | X | 200 | 0.2 | 1,570±60 | AD 690–970 |
| CAR-3 | Sabazan | GX-30425 | Shell | C. pica | Profile | XI | 230 | 2.5 | 1,460±60 | AD 790–1060 |
| CAR-4 | Sabazan | OS-41358 | Charcoal | — | Profile | X | 215 | -23.94 | 1,030±30 | AD 1290–1410 |
| CAR-13 | Sabazan | AA-67529 | Charcoal | — | Profile | XI | 53–108 | -25.6 | 988±42 | AD 980–1160 |
| CAR-14 | Sabazan | AA-67530 | Charcoal | — | Profile | XI | 53–108 | -25.6 | 1,039±35 | AD 895–1120 |
| CAR-15 | Sabazan | AA-67531 | Charcoal | — | Profile | XIII | 108–115 | -24.6 | 1,133±38 | AD 780–990 |
| CAR-16 | Sabazan | AA-67532 | Charcoal | — | Profile | XIII | 108–115 | (-25) | 1,073±38 | AD 890–1020 |
| CAR-17 | Sabazan | AA-67533 | Charcoal | — | Profile | XIV | 115–154 | (-25) | 1,172±36 | AD 770–970 |
| CAR-18 | Sabazan | AA-67534 | Charcoal | — | Profile | XIV | 115–154 | -24.6 | 1,333±57 | AD 600–780 |
| CAR-19 | Sabazan | AA-67535 | Charcoal | — | Profile | XV | 149–164 | -24.8 | 1,588±36 | AD 400–550 |
| CAR-20 | Sabazan | AA-67536 | Charcoal | — | Profile | XV | 149–164 | -25.8 | 1,584±36 | AD 410–560 |
| CAR-5 | Grand Bay | AA-62278 | Shell | C. pica | 447 | XV | 145 | 2.53 | 1,917±37 | AD 390–590 |
| CAR-6 | Grand Bay | AA-62279 | Charcoal | — | 447 | VI | 110 | -25.13 | 1,243±36 | AD 680–880 |
| CAR-7 | Grand Bay | AA-62280 | Shell | Venus sp. | 447 | VI | 127 | 3.39 | 1,789±38 | AD 530–690 |
| CAR-7 | Grand Bay | AA-62280 | Shell | Venus sp. | 447 | VI | 127 | 3.36 | 1,822±41 | AD 470–670 |
| CAR-8 | Grand Bay | AA-62281 | Charcoal | — | 447 | VI | 93 | -23.96 | 1,339±36 | AD 640–770 |
| CAR-9 | Grand Bay | AA-62282 | Charcoal | — | F016 | — | — | -25.97 | 1,227±36 | AD 690–890 |
| CAR-10 | Grand Bay | AA-62283 | Bone | Human (child—rt. fibula) | F006 | — | — | -14.21 | 1,062±44 | AD 1050–1250 |
| CAR-12 | Grand Bay | Beta-206685 | Shell | S. gigas (juvenile) | N. profile | — | 108 | 2.1 | 1,870±70 | AD 390–670 |
| CAR-11 | Harvey Vale | AA-62284 | Bone | Human (rt. ulna) | — | — | — | -12.55 | 1,027±46 | AD 1060–1280 |

Figure 8.2. Southern profile (Trench 447) at Grand Bay in July 2004 (photograph by Scott M. Fitzpatrick).

southern profile. A charcoal sample was also collected from a posthole feature (F016) and human bone from a child's burial (F006), both of which were located along the main surface eroded area and visible in the yellowish brown subsoil after clearing. The Harvey Vale sample was taken from an adult skeleton. According to eyewitnesses, the burial was found at a depth of about two meters in sandy soil. Because whether the skeleton was historic or pre-Columbian was unknown, a portion of the right ulna was dated to determine its age.

All samples were prepared using standard pretreatment procedures. Details of lab procedures can be found at the Web pages for NOSAMS (www.nosams.whoi.edu), Geochron Laboratories, Inc. (www.geochronlabs.com/14c.html), the University of Arizona (www.physics.arizona.edu/ams/), and Beta Analytic, Inc. (www.betaanalytic.com). All samples were calibrated at 2 using CALIB 5.0 (Stuiver and Reimer 1993; Stuiver et al. 1998).[1]

## Results

The suite of twenty-one radiocarbon dates from Carriacou suggests that Carriacou was first settled by ceramic-making peoples during the Terminal Saladoid period sometime between circa AD 390 and AD 500. Subsequent dates span from AD 470 to AD 1410, with most dates falling within the Troumassan Troumassoid subseries of ceramic styles cal AD 500–1000, generally correlating with our analysis of the artifact assemblages island-wide (table 8.1; figure 8.3; see Petersen et al. 2004 for discussion of ceramic classification and typology).

Chronologically, the dates from stratified deposits at Sabazan and Grand Bay represent different and consecutive periods of occupation during the later Ceramic Age, many of which overlap statistically at 2σ, an expected outcome given their context within complex midden layers (see figure 8.2). The posthole at Grand Bay dates to cal AD 690–890 and is related to three others, which are aligned 15 meters across the site in an east–west

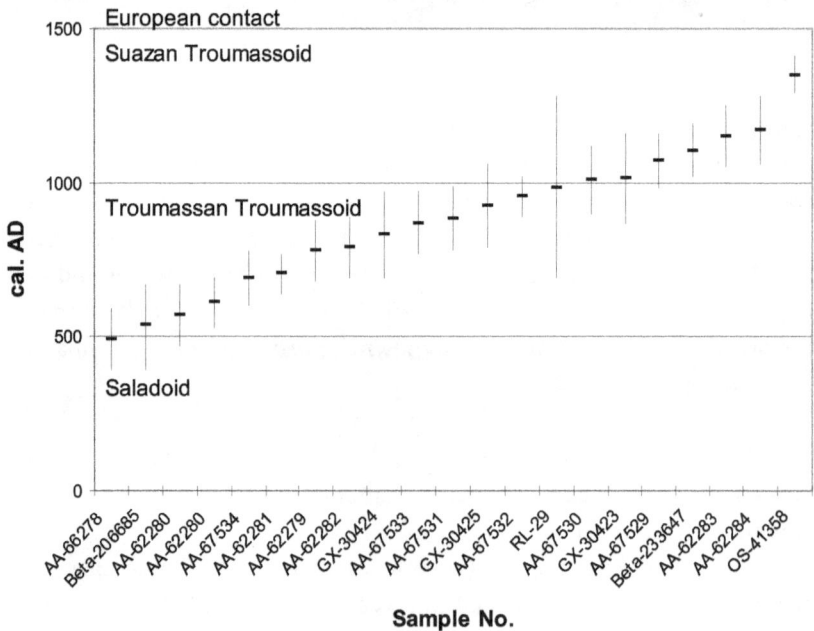

Figure 8.3. Calibrated age ranges (2σ) for the Sabazan, Grand Bay, and Harvey Vale sites.

direction. This feature appears to be the remnant of a large longhouse capable of holding several families or extended groups and is one of several possible longhouse structures at the site. The skeletons from Grand Bay and Harvey Vale date between AD 1050 and 1280; along with a charcoal sample (OS-41358), these are representative of the latest period of occupation found thus far on Carriacou, dating no later than about AD 1400. The [14]C date from the skeleton recovered by CHS museum staff at Harvey Vale also establishes that it is from a pre-Columbian individual, allowing for cross-comparisons to be made between this and several human burials found at Grand Bay.

Overall, at least sixteen of the twenty-one dates from Carriacou fall within the typically accepted time range for the Troumassan Troumassoid subseries (AD 500–1000), suggesting that this was the most intensive period of occupation on the island. However, this is also partly an artifact of sampling, and a number of dates in the lower and upper ranges for this period might be later Saladoid or early Suazan Troumassoid. Nonetheless, multiple [14]C dates taken from two distinct and exposed strata along the coast demonstrate the necessity of submitting multiple samples from complex midden deposits to confirm whether they are indeed contemporaneous.

## Discussion and Conclusions

Ongoing archaeological investigations on the island of Carriacou in the Grenadines have revealed some of the largest and archaeologically richest sites in the southern Caribbean. Results thus far indicate that the island was settled during the later part of the Saladoid period. Although ceramics such as ZIC, a stylistic feature characteristic of Early Cedrosan Saladoid (500/400 BC–AD 300/400) (see Petersen et al. 2004) were found on Carriacou, there are no definitive [14]C dates to support a pre–AD 400 occupation despite repeated attempts to collect samples from the earliest strata observed in clearly defined coastal profiles at Grand Bay and Sabazan that contain Saladoid ceramics.

One of several hypotheses may help explain this phenomenon: (1) the samples submitted for [14]C dating do not accurately represent the context from which they were collected or were somehow contaminated prior to or after analysis; (2) the island was initially bypassed by prehistoric settlers who instead decided to settle other larger islands nearby (such as Grenada, St. Vincent, and Barbados) and chose to settle Carriacou much later; (3)

Carriacou was occupied earlier than the radiocarbon dates indicate, but the evidence remains elusive; or (4) the dates accurately reflect prehistoric occupation on the island around AD 400–500, and some of the stylistically distinct ceramic motifs, which are known to occur during the earlier stages of Saladoid settlement, persisted in use later in time either as keepsakes or as part of cultural tradition.

Of these four hypotheses, the first seems unlikely—great care was taken to prevent contamination during collection and preparation prior to submission. Samples from the profiles of both Grand Bay and Sabazan were taken from freshly exposed deposits in situ and adjacent to midden deposits that contained Saladoid or other typologically distinct ceramic sherds. However, the site of Grand Bay has undergone dramatic changes due to erosion (Fitzpatrick et al. 2006), and the majority of earlier ceramic remains may have disappeared, leaving only minimal traces of Saladoid material (artifactual or otherwise).

The second hypothesis is possible, but given that Carriacou is relatively large and has abundant sources of freshwater and marine resources, as evidenced by wells found during survey (Kaye 2003) and a substantial number and diversity of faunal remains found in later archaeological deposits (LeFebvre 2007), this also seems improbable. Could settlers have initially bypassed the smaller islands in favor of larger ones, only to settle them later as populations grew?

The third hypothesis is conceivable, but why would the two largest sites on Carriacou (both of which have what would typically be interpreted as Saladoid pottery) not contain chronological evidence for this earlier settlement? Could the admixture of midden deposits through transitional periods of occupation be masking the evidence? Probably not, if dates from the two sections of the profile, separated by over one hundred meters, are contemporaneous. Local landowners have noted that the site extended out by as much as twenty to thirty meters only a few decades ago; perhaps the main Saladoid settlement was not as far inland and has since eroded away.

The fourth option seems the most likely given the current evidence. It is quite possible that Saladoid decorative traditions, some of which are more common in the earlier stages of Saladoid occupation elsewhere, persisted for hundreds of years. Three ceramic inhaling tubes (two unprovenienced from museum collections and one recovered from Grand Bay) were recently dated using thermoluminescence (TL) and optically stimulated luminescence (OSL). These samples had a weighted average of 400±189 BC, several hundred years earlier than the existing radiocarbon chronology. Although two of these samples were not recovered archaeologically, the third, dating to 190±345 BC, was found in midden deposits at Grand Bay that dated between AD 500 and 1000 and would not fall within the statistical range of the radiocarbon dates. Two additional stylistically distinct Suazan ceramic sherds dated using luminescence fell within the accepted radiocarbon chronology, and all samples analyzed petrographically appear to be made from nonlocal materials. This suggests that the inhaling bowl fragments were heirlooms transported from another island (Fitzpatrick et al. 2009). Overall [14]C and luminescence dating of various materials from stratified deposits, in a number of different contexts on Carriacou, have provided the first good cultural chronology for the island and established a strong presence for an intensive post-Saladoid occupation.

Although it is possible that an earlier Saladoid settlement was established on Carriacou but has not yet been found, or that the island was settled later in time while larger nearby islands were occupied, these seem far less likely scenarios given the evidence collected thus far. Instead, radiocarbon dates from Carriacou may lend support to the southward-route hypothesis in the colonization of the eastern Caribbean by Saladoid groups. This will need to be substantiated by further research and intensive dating regimes here and on other islands in the southern Lesser Antilles.

The appearance of Saladoid ceramics, which have long been associated with archaeological sites in the Antilles dating somewhere between circa 500 BC and AD 600, suggests that an early settlement occurred on Carriacou during this time. But why does the current suite of radiocarbon dates not support this early of an occupation? In general, there is a paucity of [14]C dates for sites and islands throughout the Caribbean and little discussion of whether these dates are actually acceptable (see chapter 2, this volume; Fitzpatrick 2006), which may help to explain some of the trends observed in patterns of Caribbean prehistoric settlement. This suggests that ranges for different cultural periods throughout the region, some of which are largely dependent on a scattering of dates and corresponding ceramic types, should be reviewed more closely for consistency and possible discrepancies.

Future investigations may indeed reveal [14]C dates that fall in line with a northward stepping-stone migration or are contemporaneous with those in the more northern islands such as Puerto Rico. But until then, archaeologists should begin submitting a greater number of samples for dating from both discrete strata and sites throughout the southern Antilles. This will not only help us to build better chronologies for this part of the Caribbean but also allow us to more effectively test hypotheses about prehistoric migration, settlement patterns, and subsequent periods of cultural development over the past six millennia.

## Acknowledgments

We thank the Grenada Ministry of Tourism, the Carriacou Tourism Office, and the local landowners at Grand Bay for granting us permission to conduct survey and excavation. Staff from the Carriacou Historical Society Museum have been extremely generous with their time and supported the field project in many different capacities, as have dozens of undergraduate and graduate students. Thanks go to Co-director Quetta Kaye and Richard Callaghan, William F. Keegan, and Corinne Hofman, who all gave useful comments on previous drafts of this paper. The radiocarbon dating of samples submitted to Geochron Laboratories, Inc., was made possible by a 2003 Research Award to Fitzpatrick. Funding for AMS radiocarbon dates was enhanced by NSF sponsorship, grants from the Quaternary Research Center and Department of Anthropology at the University of Washington, and cooperative agreements with the University of Arizona AMS Facility.

## Notes

1. Few attempts have been made to determine local marine reservoir effects in the Caribbean. Studies in the Cariaco Basin, Venezuela (Hughen et al. 1996), Jamaica, and the Bahamas (Broecker and Olsen 1961) suggest, however, that the Delta-R may be minimal. The regional average for the Caribbean is now estimated to be around -19±23, although this may change as further studies are conducted (also see www.qub.ac.uk/arcpal/marine).

# 9

## Coastal Waves and Island Hopping

### A Genetic View of Caribbean Prehistory and New World Colonization

THEODORE G. SCHURR

### Abstract

It remains unclear as to exactly when the people encountered by Christopher Columbus in 1492 migrated to the Caribbean. Archaeological evidence suggests that human groups first expanded into the American continents between fifteen thousand and twenty thousand years ago and possibly used a coastal migration route to spread throughout the New World. They apparently did not reach the Caribbean Sea until after settling much of the mainland areas of North, Central, and South America, probably between 3500 and 4000 BC. The first inhabitants of the Caribbean may possibly have come from the Yucatán Peninsula or from other areas of Central America, although some researchers suggest that the Florida peninsula could have also been a migration path. In any case, these Archaic populations were eventually replaced by Arawak groups around AD 650 and later by Carib groups, both of whom expanded north from northern South America into the Antilles. The enormous reduction of the islands' populations shortly after Spanish conquest and the subsequent repopulation of these areas by African slaves and Europeans have largely erased the indigenous biological and cultural roots of native Amerindians. These events have made the reconstruction of the settlement history of the Caribbean problematic, as this region has lost more of its aboriginal character than any other region in the Americas. Nevertheless, this chapter attempts to summarize the available evidence for the prehistory of the Caribbean and place the colonization of this region in the context of broader issues relating to the peopling of the Americas.

## Resumen

Sigue siendo confuso en exactamente cuando la gente hallada por Cristóbal Colón en 1492 emigró al área ahora conocida como el Caribe. La evidencia arqueológica sugiere que los primeros pobladores se desplazaron en los continentes americanos entre 15.000–20.000 años y posiblemente utilizaron una ruta costera para dispersarse a través del mundo nuevo. No aparenta que alcanzaron la región del Caribe hasta después de haber colonizado la mayor parte del continente del norte, central y Suramérica, probablemente entre 3500–4000 BCE. Es probable que los primeros habitantes del Caribe pudieran haber venido de la península de Yucatán o de otras áreas de América Central, aunque otros han sugerido que la península de la Florida pudo haber sido una trayectoria de la migración. En cualquier caso, estos grupos arcaicos fueron reemplazados por los grupos Caribe, que eventualmente fueron suplantados por los grupos Arawak alrededor del ANUNCIO 650, quienes se desplazaron al norte de Suramérica norteña a través de las Antillas y de las islas de Indias del oeste. La reducción enorme de las poblaciones de las islas poco después de la conquista española, a través de la esclavitud Africana, han borrado en gran parte las raíces biológicas y culturales indígenas de sus habitantes originales. Estos acontecimientos han dificultado la reconstrucción de la historia del Caribe, pues esta región ha perdido más de su carácter aborigen que cualquier otra región de las Américas. Este papel intentará resumir la evidencia disponible de la prehistoria del Caribe, y contextualizar la colonización de esta región en relación a un contexto más amplio del poblamiento de las Américas.

## Introduction

The peopling of the New World continues to be an issue of great interest both within American anthropology and outside of the discipline. Most investigations on New World colonization have attempted to delineate the larger dimensions of population movement. More specifically, they have attempted to answer the following questions: When did humans first arrive in the Americas? How many times did human groups migrate to the New World? From where did ancestral Native American populations originate? Which route(s) did ancestral Native Americans take to reach the New World?

Until recently, the view of this colonization process involved big-game-hunting populations using Clovis lithic technology that moved through an ice-free corridor in northern North America some thirteen thousand years ago before eventually dispersing throughout the American continents (Haynes 1993; Meltzer 1997; Roosevelt et al. 2002). However, as a consequence of considerable archaeological and genetic research into these questions over the past decade, our understanding of Native American history has changed quite dramatically. New archaeological data from both North and South America have overturned the previously dominant "Clovis First" model for New World colonization. Similarly, geological and climatic data from across North America have suggested that the first Americans may have used a coastal rather than an interior route in the process of colonizing the American continents. In addition, genetic data from native Siberian and American populations have provided insights into the number and timing of the migrations that reached the New World. Likewise, these data have revealed important new details about the temporal and demographic features of the colonization process.

Somewhat less attention has been given to elucidating what happened after ancestral Native Americans began expanding into the New World. Ongoing research supports the early emergence of regional cultural areas and gene pools in different parts of the North and South American continents, as well as considerable movement and contact between prehistoric Native American communities in these areas. It is this particular aspect of the colonization history of the Americas that is of greatest relevance to the Caribbean region, since it was likely settled some five thousand to six thousand years ago, well after the major continental regions were occupied (Keegan 1994, 2000).

As shown in many contributions to this volume, archaeologists working in this region are gaining a much clearer sense of the colonization history and cultural diversity of prehistoric Caribbean populations. However, there is still much to learn about the origins of these groups from a biological perspective. In this chapter, I will delve into the latter issue by exploring what we know about the prehistoric settlement of the Caribbean from a genetic perspective.

## The Peopling of the Americas: The Big Picture

After many years of research with indigenous populations in the Americas, our understanding of the pattern of genetic variation in this region of the world is becoming increasingly clear. This pattern will be outlined here through a discussion of mitochondrial DNA (mtDNA) diversity in Native American populations. Aside from its utility for molecular anthropology studies, mtDNA diversity will be examined because there are more mitochondrial sequence data from Native American and Caribbean populations than any other kind.

Although the mtDNA represents a very small proportion (16,569 base pairs) of our total genome ($3 \times 10^9$ base pairs), it contains valuable information about human movements across geography and time. Being maternally inherited, it reveals aspects of maternal genetic ancestry in human populations. Its rapid evolutionary rate also produces informative mutations or genetic markers that define different genetic lineages in these populations, with these lineages being used to track these movements in different geographic regions (Wallace et al. 1999).

Various studies of mtDNA variation in Native American populations have shown them to have five founding maternal lineages, or haplogroups, designated A, B, C, D, and X (Schurr et al. 1990; Torroni et al. 1992, 1993; Forster et al. 1996; also see references in Schurr 2004) (figure 9.1). These lineages are differentially distributed throughout the Americas, with haplogroups A and B tending to occur at higher frequencies in North and Central America, and haplogroups C and D at higher frequencies in South America. Within these continental areas, we can see regional patterns of haplogroup frequencies that probably reflect both the initial phase of colonization and the population dynamics (expansions, gene flow, founder effects) that occurred over the next fifteen thousand years (Crawford 1998; Salzano and Callegari-Jacques 1988; Salzano 2002; Schurr 2004). In this respect, it is worth noting that there are few, if any, data for Caribbean Indian populations, leaving a large lacuna in the distribution of mtDNA haplogroup frequencies in the Americas (figure 9.1). Not surprisingly, the general absence of genetic data from this region complicates efforts to reconstruct the prehistoric settlement of the Caribbean.

Along with giving some resolution to the pattern of genetic diversity in contemporary Native American groups, mtDNA data have also been used to estimate the timing of the initial entry of the founding populations.

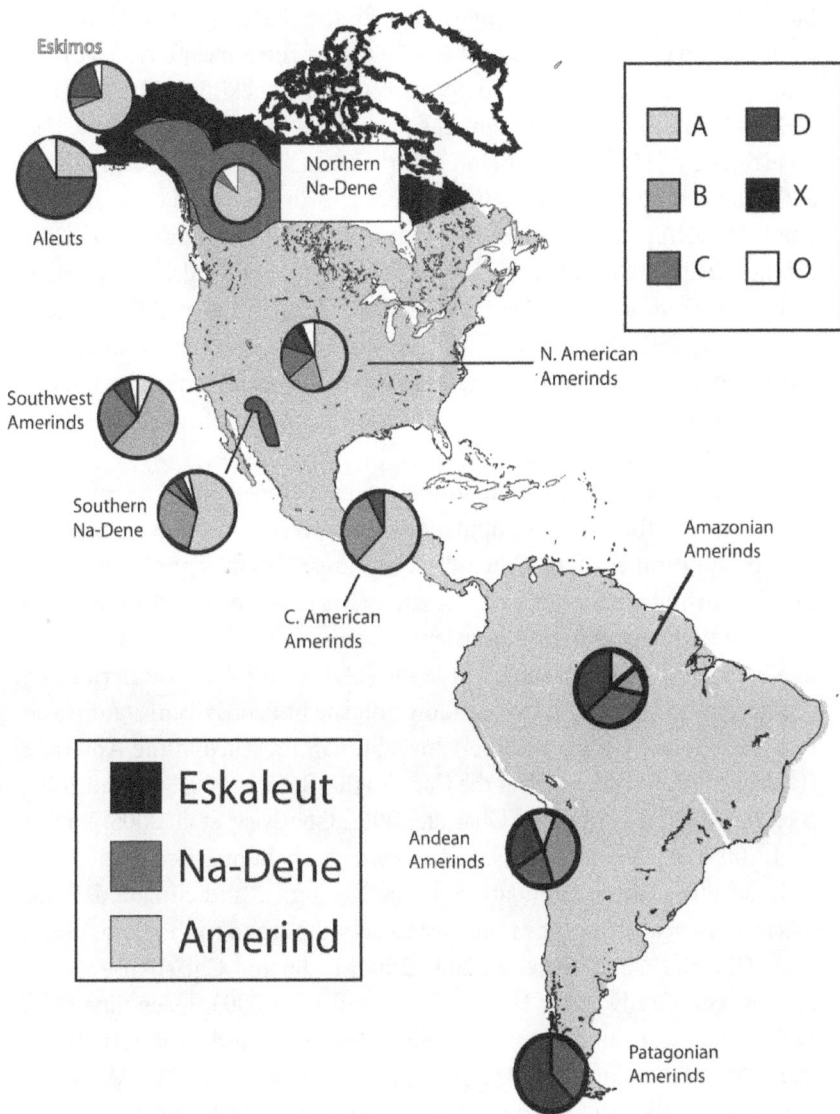

Figure 9.1. Map of mtDNA frequency in the Americas.

There have been various attempts to date the antiquity of the five mtDNA haplogroups present in the Americas, with the range of ages obtained by these studies between 13,000 and 40,000 years (Torroni et al. 1993; Forster et al. 1996; Bonatto and Salzano 1997; Stone and Stoneking 1998; Silva et al. 2002). Our recent effort to refine the date of the initial entry time yielded an estimate of 16,275 years before the present (YBP) (Schurr and Sherry 2004), and this date is consistent with recent estimates based on analysis of whole mtDNA genome sequences (Tamm et al. 2007; Fagundes et al. 2008; Kitchen et al. 2008; Perego et al. 2009). Similar dates were also estimated in recent Y-chromosome (Seielstad et al. 2003; Bortolini et al. 2003; Zegura et al. 2004) and multilocus (Hey 2005) analyses. Together, these genetic coalescence or divergence estimates imply that the first Americans began expanding in the New World prior to the emergence of the Clovis lithic culture (13,020 cal BP).

In parallel with these genetic studies, recent archaeological research has also suggested that human populations were present in the Americas before the appearance of the Clovis lithic cultures. This interpretation comes from a growing number of archaeological sites from both North and South America that appear to be equivalent in age or older than the Clovis sites in North America. The Monte Verde site (14,865 cal BP) in southern Chile (Dillehay 1997, 1999) and the Meadowcroft site in Pennsylvania (Adovasio et al. 1998) clearly suggest a pre-Clovis human presence in the Americas (Dillehay 1999), while work at the Cactus Hill (Bower 2000; Stokstad 2000; McAvoy 2005) and Topper (Chandler 2001; Goodyear et al. 2005; Waters et al. 2005) sites has also suggested a similar early human occupation.

In addition, the accumulation of new geological and climate data for North America dating from the past twenty thousand to thirty thousand years (Burns 1996; Fedje et al. 2001, 2004; Fedje and Christiansen 1999; Heaton and Grady 2003; Hetherington and Reid 2003; Josenhans et al. 1997; Mandryk et al. 2001) is providing a revised picture of the pattern of glaciation that occurred during the Last Glacial Maximum (LGM). These data suggest that the ice-free corridor once thought to be the conduit through which ancestral Native Americans moved from Siberia into the Americas may not have been available for human use until twelve thousand to thirteen thousand years ago. In other words, this corridor would not have been sufficiently accessible until right around or shortly after the appearance of the Clovis cultures in North America.

Therefore, given the archaeological visibility of human groups in the

Americas by at least 15,000 YBP, the initial colonizers would appear to have used a different route to the reach the portion of the North American continent not covered by ice. The most plausible path for ancestral Native Americans to have taken is a coastal route along the western edge of northern North America (Fladmark 1979; Dixon 2001, 2002; Heaton and Grady 2003). The presence of ice-free refugia would have allowed populations moving by watercraft to traverse the southern edge of the Beringian land mass, temporarily land at these refugia, and leapfrog around the ice sheets until reaching unglaciated areas to the south.

In a recent effort to model this colonization process, Fix (2005) analyzed published mtDNA data from different regions of North America to determine which kind of migration scenario would explain the current haplogroup frequencies in modern Native American populations. The results of his analysis clearly suggested that founding Native American populations followed coastlines to expand into different parts of North America, including the marine coasts and the major river systems that extended into the interior of the continent (figure 9.2), and possibly colonized the entire coast of North America in no more than three thousand years. Likewise, human groups would have moved along coastal areas to colonize the South American continent at roughly the same time, perhaps in a bipartite fashion, as suggested by other genetic research (Salzano and Callegari-Jacques 1988; Mesa et al. 2000; Keyeux et al. 2002). The pioneering coastal settlements in these marine and riverine environments would have served as "staging areas" for the continuing colonization of interior regions (Anderson and Gilliam 2000).

Ancient DNA (aDNA) studies have also revealed important aspects of the colonization of the Americas, both the initial phase of expansion and the subsequent settlement of human groups in different regions. The later phase of the peopling process is reflected in the continuity of mtDNA haplogroup frequencies between ancient and modern populations from certain regions of North American (O'Rourke et al. 2000a, 2000b; Malhi et al. 2002). However, there are some exceptions to this trend, with these discontinuities likely occurring because of climatic change, the intrusion of outside groups, or other reasons (Kaestle and Smith 2001). The general regional continuity of Native American populations over the past four thousand to six thousand years further suggests that the Caribbean region might also exhibit this kind of pattern, should enough data from ancient and modern populations be available for comparison.

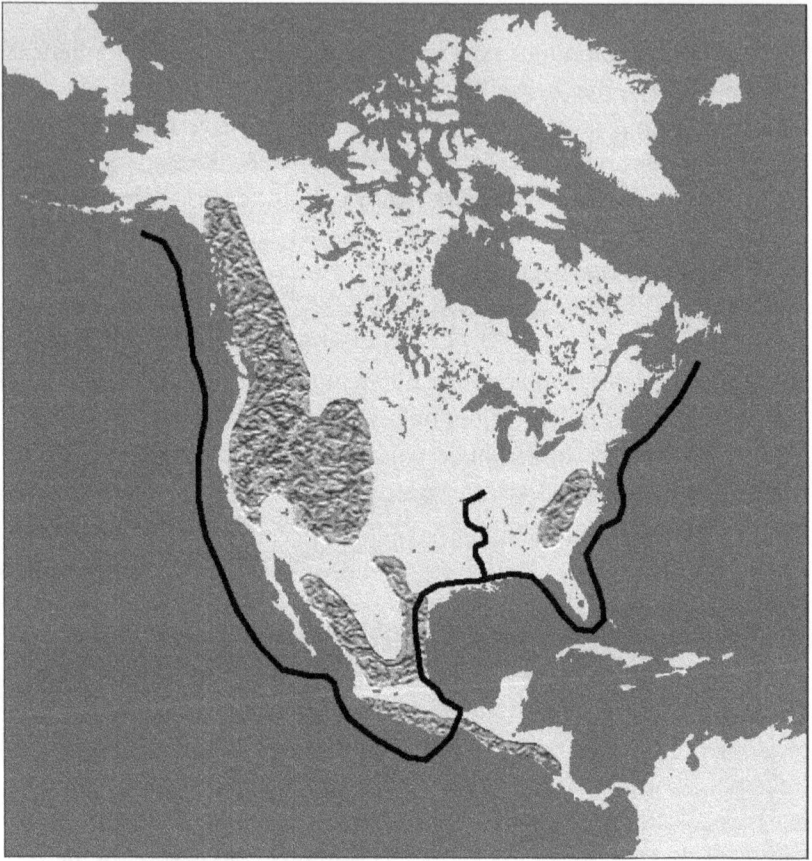

Figure 9.2. Coastal colonization route.

Such aDNA studies are also illuminating aspects of the demographic history of Native American populations. Figure 9.3 presents a network of mtDNA haplotypes belonging to haplogroup A that come from ancient and modern Native American individuals (Malhi et al. 2002). In this network, one observes that modern populations do not contain all of the haplotypes that were once present in Native American groups. Through stochastic processes such as genetic drift and founder effects and because of the demographic crash in Indian populations resulting from European contact (Thornton and Marsh-Thornton 1981; Crawford 1998; Ubelaker 1988; Mann 2005), a number of haplotypes that were once present in Native American groups have not persisted into the present. For this reason, many of the aDNA haplotypes appear at intermediate nodes or on

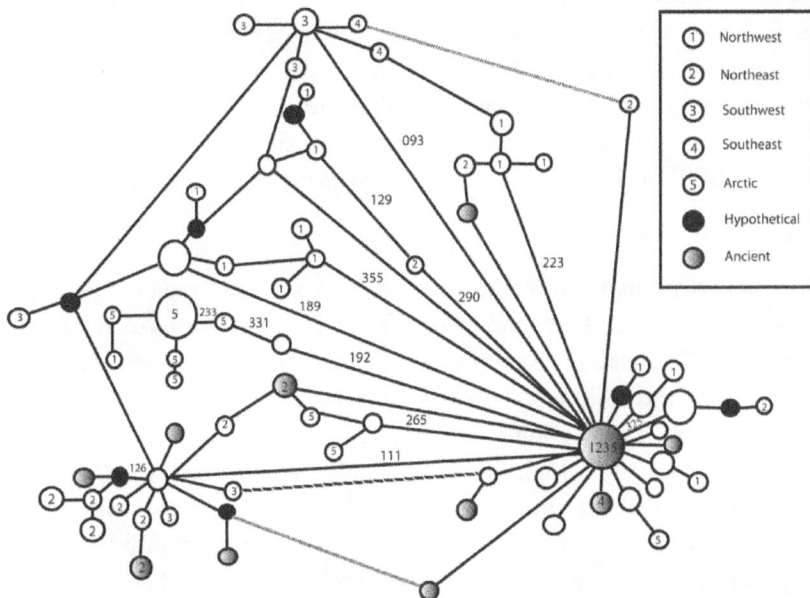

Figure 9.3. Network of haplogroup A sequences.

unique branches in the network compared to those from modern Native American populations. Therefore, by indicating the nature of diversity in prehistoric populations, and by revealing the degree of similarity of past populations to their modern descendants, the aDNA data provide an important diachronic perspective on genetic diversity in Native American populations.

## Genetic Diversity in the Caribbean

The preceding overview of mtDNA diversity in the Americas sets the stage for the investigation of the prehistoric settlement of the Caribbean from a genetic perspective. As will be discussed below, a number of factors hinder efforts to reconstruct the genetic prehistory of Caribbean Indians. Among these are the virtual absence of extant tribes throughout the region and a paucity of well-preserved skeletal remains that could provide aDNA for genetic analysis. DNA data from modern Caribbean populations may be of some use for this reconstruction effort but are complicated by historical admixture with both European colonizers and Africans brought to the region through the trans-Atlantic slave trade (for example, see Keegan 1992, 1996). As a result, one must glean the important details about the colonization process from a relatively limited data set. For this reason, the information provided by archaeological, linguistic, and osteological studies of prehistoric populations (as discussed in the other chapters of this volume) will continue to be central to studies of Caribbean prehistory.

Caribbean Colonization Models

As amply demonstrated by this volume, the Caribbean region is defined by a series of islands that connect North, Central, and South America (see figure 0.1). The relative proximity of these islands to each other, and their closeness to the continental areas of the Americas, likely facilitated the movement of human populations into and around the Caribbean basin once it was initially colonized (Keegan 2000; see Petersen et al. 2004 for a recent review of the chronological schemes for early settlement). Such movements also undoubtedly involved the use of watercraft by colonizing groups. In this regard, if the Caribbean basin is viewed as a separate geographic entity like the North American continent, then the major questions concerning its prehistory are much the same as for the Americas as a whole: When did human groups first enter this region? How did they spread through it? How many expansions account for the archaeological, linguistic, and biological diversity in Caribbean Indians? What happened once human groups began settling on the islands of this region?

There are several alternative hypothetical migration routes that ancestral Native American populations could have taken to reach the Caribbean islands (figure 9.4). These include putative movements (A) across the Straits of Florida from North America into Cuba and then dispersing eastward; (B) from Central America, across the Yucatán passage, to Cuba; and (C) from northern South America (Venezuela), followed by dispersal into the Lesser Antilles and westward (Rouse 1992a; Keegan 1992, 2000; Moreira de Lima 1999). The latter Antillean dispersal model is currently favored with respect to the expansion of Carib and Arawakan tribes in the Caribbean over the past two thousand years but may not explain the initial settlement of this region.

In principle, it would be possible to test these three hypotheses by examining genetic variation in populations from these putative source areas and those living on the Caribbean islands. However, because there are no living Caribbean Indian populations, one must analyze genetic variation

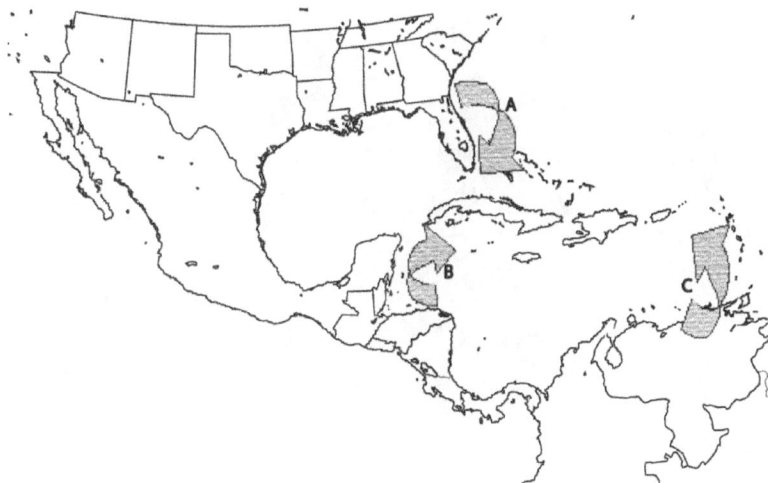

Figure 9.4. Hypothetical migration routes.

in the contemporary populations of these islands, however admixed, to identify genetic lineages common to Native American populations in them. In addition, one needs aDNA data obtained from prehistoric skeletal remains recovered on different Caribbean islands, as well as those from archaeological populations from the surrounding continental regions, to have prehistoric reference points against which to compare the modern DNA data. It is with the aDNA data that the discussion of genetic diversity in the Caribbean will begin.

Ancient DNA Data from the Caribbean

The first efforts to genetically characterize prehistoric Caribbean peoples involved the analysis of mtDNA variation in human remains from Hispaniola (Dominican Republic). Lalueza-Fox and colleagues (2001) found that pre-Columbian Taínos from Hispaniola had reduced mtDNA diversity compared to modern Native American populations and high frequencies of haplogroups C and D. They interpreted these data as showing that the colonizing population underwent a founder effect during the colonization of the Caribbean region. They also pointed to a South American origin for the ancestral Taínos, since other Arawakan-speaking populations from northern South America also had high frequencies of these mtDNA lineages (Torroni et al. 1992, 1993; Merriwether et al. 1995; Easton et al. 1996).

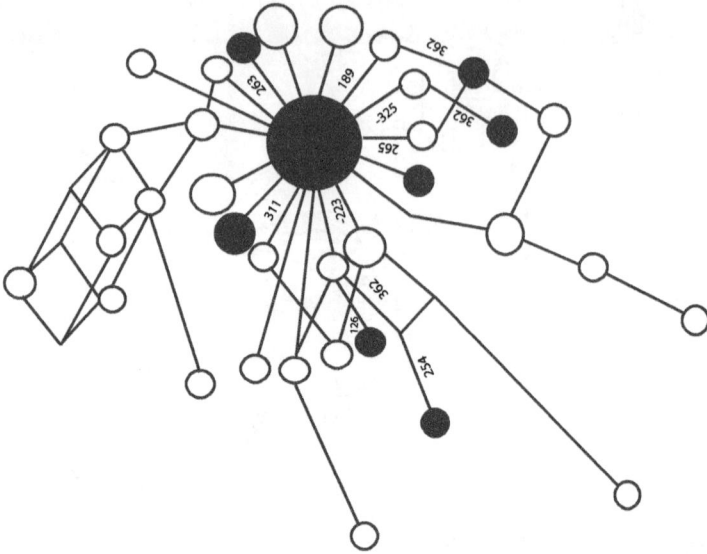

Figure 9.5. Haplogroup C network for the Taínos.

Figure 9.5 presents a reduced median network of mtDNA haplotypes from haplogroup C that includes data from the prehistoric Taínos and modern South American Indians. It shows that the Taíno haplotypes represent a small subset of the total present in living South American tribes. A few of these haplotypes appear to be unique to the Taínos, while others belong to branches on which modern haplotypes can also be found. The central haplotype has a sequence motif (16223–16298–16325–16327) that is commonly seen in South American Indians (Torroni et al. 1993; Merriwether et al. 1995; Easton et al. 1996; Santos, Ribeiro-Dos-Santos, et al. 1996; Ward et al. 1996; Mesa et al. 2000; Bert et al. 2001), although also present in North American populations (Torroni et al. 1993; Lorenz and Smith 1997; Malhi et al. 2001, 2002, 2003; Bolnick and Smith 2003).

Lalueza-Fox and colleagues (2001) also calculated genetic distances from the haplogroup frequencies in the ancient Taínos and those in a variety of modern tribes from Central and South America, then examined the population affinities of these tribes through principal components (PC) analysis (figure 9.6). As seen in the PC plot, the Taínos clustered most closely with the Yanomami and then to other tribes present in Amazonian South America but were not close to Central American populations, such as the Huetar, Ngöbé, and Kuna.

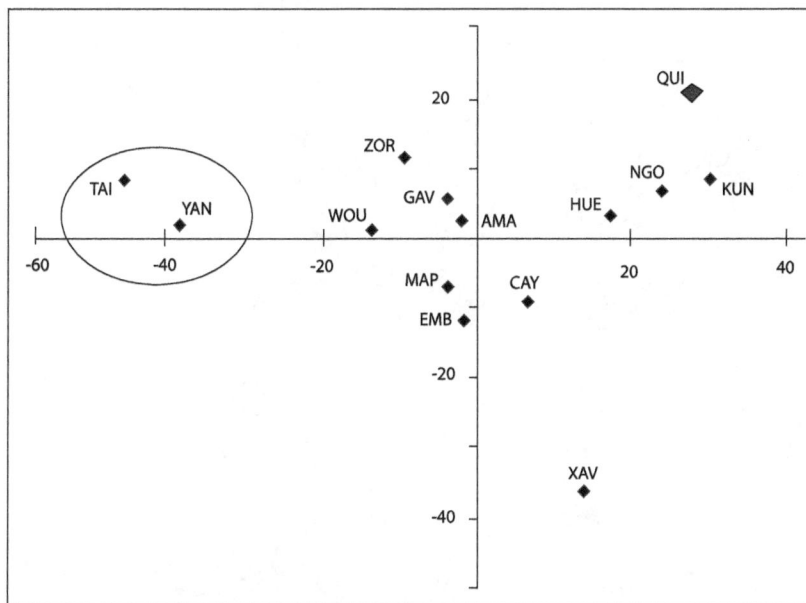

Figure 9.6. Principal components analysis plot for Amerindians.

However, their later study of pre-Columbian Ciboneys from Cuba yielded results that could not rule out Central America as a possible source area (Lalueza-Fox et al. 2003). As for the Taínos, haplogroups C and D comprised the majority of their mtDNAs. However, in the reduced median network generated from the mtDNA data, most Ciboney haplotypes from haplogroup C were distinct from those in the Taínos, although two haplotypes from each population belonged to the same sublineage (figure 9.7). In addition, several of the Taíno haplotypes appeared on branches containing other haplotypes seen in Central American and/South American tribes. Interestingly, none of the Ciboney or Taíno haplotypes appeared to show close similarities to those from North American tribes, implying that they did not come from founding populations who used the Florida Straits migration route.

Although these results are very interesting, the nature of aDNA data places limits on the kinds of comparisons between genetic patterns in ancient and modern Caribbean populations that one can make. At many archaeological sites, aDNA samples are obtained from human remains that are hundreds to thousands of years apart in age. For example, the Ciboney samples analyzed by Lalueza-Fox and colleagues (2003) came from three

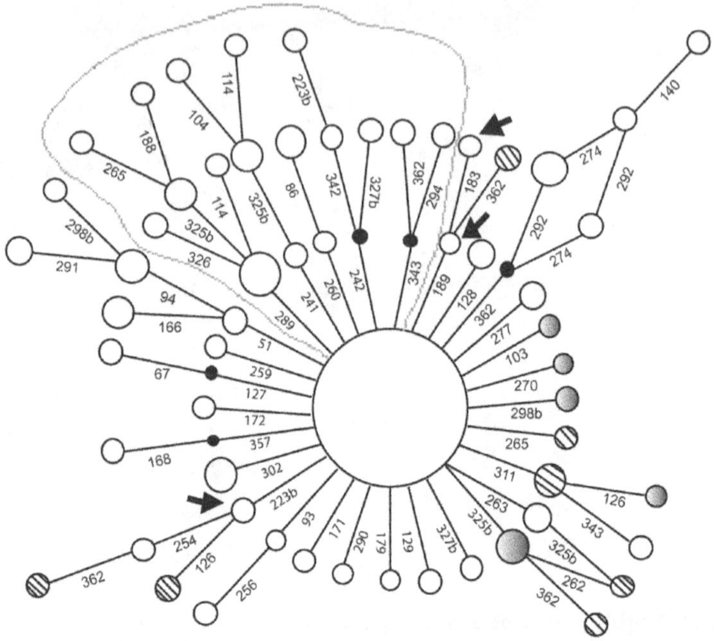

Figure 9.7. Haplogroup C network for the Ciboneys.

archaeological sites dating between 1620 BP and 4700 BP. Consequently, they cannot necessarily be considered to belong to a single population or to represent a synchronic snapshot of genetic diversity on a particular island (Cuba). In addition, the limited number of human remains recovered on Caribbean islands, particularly from older sites, may not provide a statistically valid sample size that would allow strong inferences to be drawn about the patterns of genetic diversity observed in them. For example, in the previously described aDNA studies, only fourteen Taíno and fifteen Ciboney samples were characterized for mtDNA variation (Lalueza-Fox et al. 2001, 2003). Therefore, such data can provide only a glimpse of the overall genetic diversity that was present in a particular location at a particular moment in time.

Historical Genetic Admixture in the Caribbean

Aside from the limited genetic information obtained from ancient populations, there are other factors that complicate efforts to reconstruct Caribbean prehistory solely on genetic evidence. The first is the remaking of the

Caribbean gene pool during the past five hundred years. This process began with the arrival of Columbus at Hispaniola in 1492 and continued over the next several centuries. As noted above, with European contact came demographic losses in native populations resulting from warfare, disease, and slavery. The subsequent introduction of African slaves for forced labor also led to significant admixture between Indians and Africans, and often Africans and Europeans wholly replaced the indigenous inhabitants of an island once they disappeared from it. The extent to which Indian populations from different Caribbean islands were decimated and/or absorbed into the colonizing populations is still not entirely clear.

An example of this biogenetic transformation is the emergence of the Garifuna, or Black Caribs (figure 9.8). Over the course of three hundred years, Africans absorbed or melded with local Indian groups in the Antilles, forming new (hybrid) populations. These groups practiced the traditional lifeways of the Carib Indians but possessed different biological and phenotype characteristics than the original inhabitants of the islands (Crawford et al. 1981, 1984; Crawford 1986). When the British relocated them to the Atlantic coast of Central America, they began expanding in number and dispersing to other parts of that region. Genetic analyses of the Garifuna have shown that, in terms of mitochondrial diversity, they have a strong African profile, with mostly haplogroup L1, L2, and L3 haplotypes being present in them (Monsalve and Hagelberg 1997; Salas et al. 2005). Although based on small sample sets, these results suggest that the indigenous mtDNA contribution to the Garifuna was relatively small.

In general, Caribbean populations represent a mixture of African, European, and Native American genetic ancestries. The degree of ancestry attributable to these three source areas or populations varies from island to island, whether Cuba, Puerto Rico, the Dominican Republic, or Barbados, and often differs when measuring admixture with mtDNA (maternal) and Y-chromosome (paternal) variation, to a large extent depending on which European power controlled the island in question. On some islands, there is greater African than Native American and European maternal ancestry, while, in all locations, there is much greater European than Native American male genetic ancestry (see, for example, Torroni et al. 1995; Tajima et al. 2004).

In this regard, the Puerto Rican population exhibits a remarkable diversity of Native American mtDNAs compared to those from other islands. The average frequency of Native American haplogroups across the island

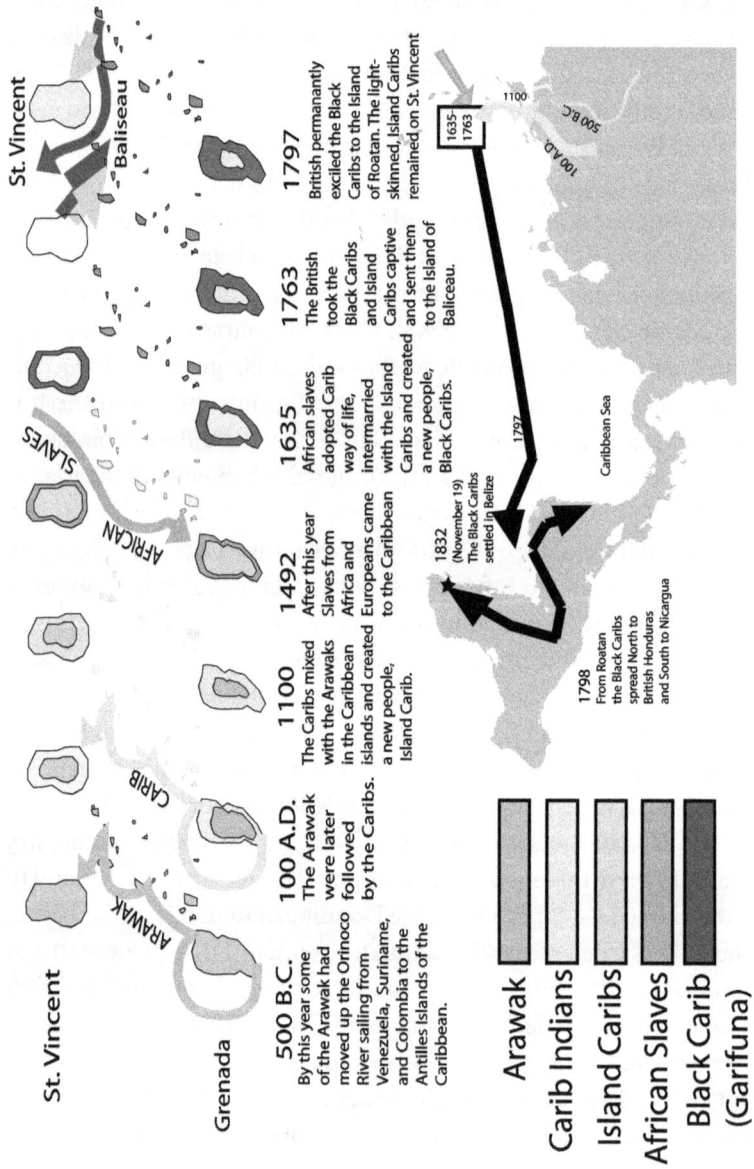

Figure 9.8. Garifuna migrations.

is around 61 percent, far higher than seen elsewhere in the Caribbean, with populations from certain local areas showing up to 80 percent indigenous haplotypes (chapter 3, this volume; Martínez-Cruzado et al. 2001, 2005). These data suggest that the indigenous populations of Puerto Rico were largely absorbed or assimilated into the colonizing populations of Europeans and Africans and that their mtDNA haplotypes have been preserved in their admixed descendants. This extensive data set will be invaluable for testing the different migration models for Caribbean colonization, especially for assessing the source area for the ancestors of the Taínos.

European Contact with and Removal of Southeastern Indians

Another historical event that could affect efforts to reconstruct the prehistoric settlement of the Caribbean is the population loss suffered by Indian tribes in the southeastern United States (in particular, Florida) as a consequence of European (Spanish) exploration or conquest of the region. De Soto's expedition through this region was enormously disruptive for the tribes living there, with warfare and disease killing a significant number of individuals from indigenous communities during and after his journey (Clayton et al. 1993; Galloway 1997; Hudson 1997). There were similar demographic consequences for native populations who came into contact with Europeans elsewhere in North America (Mann 2005). In addition, the Indian Removal of 1814–58 led to the deaths of thousands of people belonging to the Cherokee, Creek, Choctaw, Chickasaw, and Seminole tribes and the ultimate relocation of these tribes to western reservations (Foreman 1953; McDonnell 1991). For these reasons, it may not be possible to identify which of these tribes or their ancestors might have been involved in the initial colonization of the Caribbean from the north. Therefore, one must extrapolate the genetic composition of the ancestral groups who lived in the region at that time from the patterns of genetic diversity (that is, mtDNA haplogroup frequencies) in modern Indian populations that originated in the southeastern United States.

## Implications of Genetic Data from the Caribbean

In spite of the limitations of the DNA evidence from the Caribbean, one can still examine the genetic diversity within populations inhabiting this region in an effort to find connections between them. Figure 9.9 presents mtDNA haplogroup frequency data for both ancient and modern

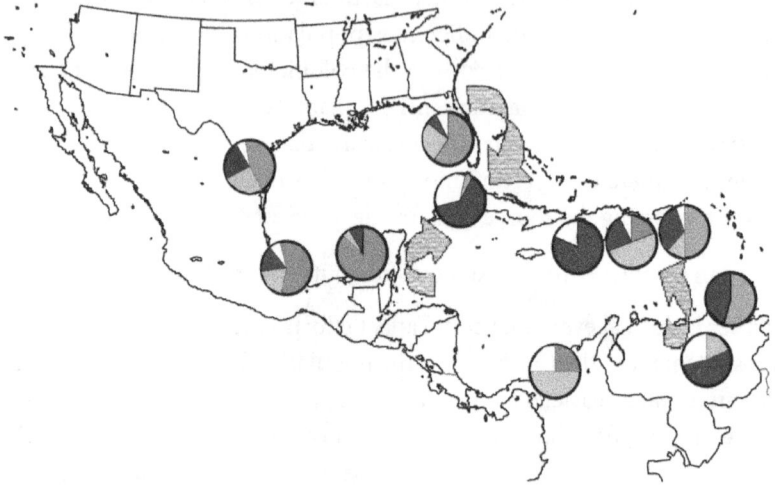

Figure 9.9. Map of Caribbean mtDNA haplogroup frequencies.

populations from around the Caribbean basin. Although these data should not be seen as providing a comprehensive picture of genetic diversity there, one can assess their implications for the colonization history of the region, as well as the evolutionary and historical forces that might have shaped them.

To begin with, the older Ciboneys show a predominance of haplogroups C and D, as also noted in the Taínos (Lalueza-Fox et al. 2001, 2003). The Arawakan-speaking Yanomami also show high frequencies of these maternal lineages, as do other populations from the Amazon region (Torroni et al. 1993; Santos, Ribeiro-Dos-Santos, et al. 1996; Ward et al. 1996; Mesa et al. 2000; Bert et al. 2001), although they also have haplogroups A and B (Torroni et al. 1993; Merriwether et al. 1995; Easton et al. 1996). Only the prehistoric Ciboneys exhibit any haplogroup A mtDNAs. These findings continue to suggest that Arawakan speakers from South American were the source populations for the Arawak and Carib populations of the Caribbean (Rouse 1992a; Keegan 1992, 1996; Wilson 1997).

By contrast, Mexican Indians have high frequencies of haplogroup A and B mtDNAs, with lower frequencies of C and negligible frequencies of

D. This pattern holds true for populations from many parts of Mexico, including Maya from the Yucatán Peninsula (Torroni et al. 1992, 1993, 1994; Green et al. 2000; Kemp et al. 2005), a potential launching point for ancient Mesoamerica groups to expand into the Caribbean islands, as well as for Mesoamerica in general (Torroni et al. 1993, 1994; Santos et al. 1994; Kolman et al. 1995; Kolman and Bermingham 1997). This is not the same mtDNA haplogroup profile as that seen in the samples from Antillean or Arawakan speakers in the Caribbean or South America. When ancient DNA data from Maya burials are included in this analysis (González-Oliver et al. 2001), one observes a slight shift from a very high frequency of haplogroup A to a moderate frequency in Maya groups over time. Such a shift in haplogroup frequencies might also have occurred in ancient Caribbean populations over the past several thousand years, giving rise to the pattern of genetic diversity observed in the aDNA samples.

From a North American standpoint, the Florida Seminoles show a haplogroup profile that is somewhat similar to that seen in Mexican tribes, with high frequencies of A and B and some C (Huoponen et al. 1997; Bolnick and Smith 2003). This pattern is also seen in the Choctaw, Chickasaw, and Cherokee samples (albeit with varying different frequencies of A, B, and C) but not for the Creeks, who have considerable frequencies of D (Merriwether et al. 1995; Bolnick and Smith 2003). Thus, there is not a uniform pattern of haplogroup frequencies for southeastern Indian tribes, although they do share some haplotypes within each mtDNA lineage (Bolnick and Smith 2003). While further review is warranted, these haplotypes do not appear to be closely related to those appearing in the ancient Taíno and Ciboney samples.

When we shift our focus south from Florida to the Caribbean itself, we see that populations from the Dominican Republic have all four major DNA haplogroups (A–D) at considerable frequencies (18–36 percent; Tajima et al. 2004). All four of these maternal lineages are also present in Puerto Ricans, with haplogroups A (32 percent) and C (22 percent) occurring most frequently on the island (Martínez-Cruzado et al. 2005). Additional analysis will be required to determine the extent to which Dominicans and Puerto Ricans share specific haplotypes within these haplogroups. By contrast, Garifuna populations have only haplogroup A and C mtDNAs (with small sample sizes), with most of their haplotypes belonging to African haplogroups (Monsalve and Hagelberg 1997; Salas et al. 2005).

In light of these data, one could ask why haplogroup A is so common in contemporary Native American groups from all around the Caribbean basin but not in the prehistoric Ciboneys and Taínos. Is this due to the loss or extinction of haplogroups A and B in the Indian populations of the Caribbean (Martin and Steadman 1999)? Did drift or founder effects eliminate these lineages from those populations? Investigating this discrepancy in patterns of genetic diversity between ancient and modern populations will be crucial for elucidating the connections between the Caribbean and continental regions of the Americas, and the population dynamics of the prehistoric Caribbean groups themselves.

## Future Directions

Despite their limitations, the ancient DNA data from prehistoric Caribbean populations are still useful for addressing questions about the number and timing of migrations and the sources of migrating populations outside of the Caribbean Sea. They allow some assessment of the biological and cultural continuity or discontinuity of prehistoric populations and perhaps also the extent of interisland travel and mate exchange. However, there may not be sufficient sample sizes from archaeological populations to obtain precise answers to these questions, largely because of poor preservation of human remains in this region.

Therefore, to expand our knowledge of Caribbean Indian ancestry and the colonization of this region, we will need broader genetic surveys of modern populations from Haiti, the Dominican Republic, Cuba, the Bahamas, and other islands. Given that all Caribbean Indian tribes are now virtually extinct, these admixed populations are the only remaining reservoirs of Native American haplotypes and alleles that can be analyzed. In addition, the accumulation of more genetic data from Mexican Indians living in Yucatán and Caribbean coastal areas would allow a more robust testing of the putative Central American source of ancestral Caribbean populations.

It will be equally important to begin synthesizing the data and inferences drawn from genetic data with those from archaeological, linguistic, and osteological studies. This synthetic approach will have considerable explanatory power because the genetic data are not currently sufficient in and of themselves to resolve the major issues about the prehistoric settlement of the Caribbean. Thus, one could compare [14]C dates from different levels at archaeological sites and coalescence times estimated for genetic lineages present in archaeological or modern populations to yield insights into the timing of island colonization. In addition, the comparison of skeletal and genetic data from the same individuals might reveal whether different osteological populations (Saladoid and Ostionoid; Rouse 1992a) actually represent distinct genetic groups or just an osteologically variable population (chapter 5, this volume; Ross et al. 2002a; Ross 2004). Such comparisons could further illuminate interisland exchanges (chapter 7, this volume), as well as the emergence of regional sequences of cultures within the Caribbean basin (chapters 2, 6, and 8, this volume).

## Acknowledgments

I would like to thank Ann Ross, Dominique Rissolo, and Deborah Thomas for their helpful discussions of issues pertaining to the history of Caribbean populations.

# Epilogue

## Linking Caribbean Shores, Visualizing the Past

SCOTT M. FITZPATRICK

As the chapters in this book illustrate, there is an increasing amount of research in a number of subfields within anthropology—namely, archaeology, skeletal biology, and genetics—that are providing exciting new results about the variables that encouraged (or perhaps discouraged) peoples to settle the Antilles over a span of several thousand years from Central and South America. The results are extremely relevant not only for Caribbeanists but also for researchers interested in broader questions relating to New World colonization, population affinities, and the development of intensive maritime economies, among many others.

As readers peruse these chapters, they will no doubt recognize that some well-known scholars such as the late Irving Rouse were highly influential figures when it came to constructing early models of pre-Columbian Caribbean culture-history. But as several authors in this book have noted, these models were structured largely on ceramic typologies based on very few or no radiocarbon dates. Subsequent research over the past few decades has continued to use these sequences of pottery to define migrations and culture change, primarily because they were, one might argue, overly convenient units of analysis. But as research presented here and elsewhere is showing more clearly, these sequences have proven to be chronologically inadequate and too simplistic. To more definitively anchor the ages of initial settlement and long-term changes that occurred in the Caribbean islands after humans first arrived, a number of archaeologists have begun focusing more on compiling lists of radiocarbon dates and their associated assemblages to examine whether these are acceptable or not, based on various criteria (such as sample type, provenience, and so forth).

In this volume, we stress the importance of using different analytical techniques to examine population movements. Although radiocarbon dating is not new per se, having been the most widely used chronometric technique in the Caribbean just as it has for archaeologists elsewhere in the world, the dates returned from labs have largely been taken at face value without any assessment of their overall validity. Three chapters in this volume (chapters 2, 4, and 8), in fact, address this central issue, which is now changing the way we view the timing of and span of colonization events, migration routes, periods of interaction, the use of ceramic typologies as a proxy for time and culture, and a host of other behaviors. By closely investigating potential sources of contamination for dates and their provenience, we can rule out those that appear problematic. Using this process of chronometric hygiene in conjunction with continually updated calibration curves, we then can adjust and refine our chronologies so that they more accurately reflect temporally distinct cultural behaviors. It would be an understatement to say that these kinds of examinations were long overdue, and I am particularly encouraged that Caribbeanists are drawing more attention to the issue.

An additional suite of advanced techniques developed by geneticists and skeletal biologists has also increased our ability to track population movements in the Caribbean. Chapters 3, 5, and 9 provide us with some excellent examples of how mitochondrial DNA, ancient DNA, and skeletal morphometrics can help us to establish population affinities within and between pre-Columbian and modern groups. The significance of these studies is not only that they are some of the few for the Caribbean to date but also that they help to bring the Caribbean into the broader discussion of New World colonization. The Caribbean has been seen for far too long as peripheral to these discussions; these chapters, and the volume as a whole, will hopefully serve to alter this perception.

We should not forget two other major contributions that are gaining wider acceptance and utility in the archaeological and anthropological communities when it comes to deciphering human population movements: computer simulations of seafaring and strontium isotope analysis. In taking the first approach, the analysis in chapter 6 is intriguing, for it looks at both oceanographic conditions and volcanism to explain why Archaic peoples may have ventured to some islands more easily than others. The contribution in chapter 7, following the second research path, is equally fascinating—because strontium isotope signatures are unique to islands, it is possible to re-create the life history of individuals and determine whether they lived their whole life on the island where they were born or moved away (and in some cases were brought back to their home island to be buried after death).

The Caribbean holds important clues as to how, when, and why peoples migrated to the islands prehistorically; each of the chapters in the volume reinforces the notion that intra- and interdisciplinary work has much to offer. I am pleased that we were able to bring together such an outstanding group of scholars to investigate these issues in the Caribbean. As we continue to explore these islands archaeologically, expand our biological and genetic research, and incorporate other sophisticated techniques into our arsenal of academic inquiry, I am confident that we will find that this region of the Americas is pivotal for understanding how Amerindians in the past structured their migrations, settled down, and interacted with one another over the course of millennia.

# References

Achilli, A., U. A. Perego, C. M. Bravi, M. D. Coble, Q. P. Kong, S. R. Woodward, A. Salas, A. Torroni, and H. J. Bandelt. 2008. The phylogeny of the four pan-American mtDNA haplogroups: Implications for evolutionary and disease studies. *PLoS ONE* 3 (3): e1764.

Adovasio, J. M., D. R. Pedler, J. Donahue, et al. 1998. Two decades of debate on Meadowcroft Rockshelter. *North American Archaeologist* 19 (4): 217–41.

Afifi, A. A., and V. Clark. 1996. *Computer-aided multivariate analysis*. 3rd edition. London: Chapman and Hall.

Alegría, R. E. 1965. On Porto Rican archaeology. *American Antiquity* 3: 246–49.

Alegría, R. E., H. B. Nicholson, and G. R. Willey. 1955. The Archaic Tradition of Puerto Rico. *American Antiquity* 21 (2): 113–21.

Allaire, L. 1990. Prehistoric Taino interaction with the Lesser Antilles: The view from Martinique, F.W.I. Paper presented at the 55th Annual Meeting of the Society for American Archaeology, Las Vegas, Nevada.

Allaire, L., and M. Mattioni. 1983. Boutbois et Le Godinot: Deux gisements acéramiques de la Martinique. In *Proceedings of the IX International Congress for the Study of the Pre-Columbian Cultures of the Lesser Antilles*, ed. L. Allaire and F.-M. Mayer, 27–38. Montreal: Centre de Recherches Caraibes.

Alves-Silva, J., M. da Silva Santos, P. E. M. Guimarães, A. C. S. Ferreira, H. J. Bandelt, S. D. J. Pena and V. Ferreira Prado. 2000. The ancestry of Brazilian mtDNA lineages. *American Journal of Human Genetics* 67 (2): 444–61.

Ammerman, A. J. 1981. Surveys and archaeological research. *Annual Review of Anthropology* 10: 63–88.

Anderson, D. G., and J. C. Gillam. 2000. Paleoindian colonization of the Americas: Implications from an examination of physiography, demography, and artifact distribution. *American Antiquity* 65: 43–66.

Anderson, S., A. T. Bankier, B. G. Barrell, M. H. L. de Bruijn, A. R. Coulson, J. Drouin, I. C. Eperon, et al. 1981. Sequence and organization of the human mitochondrial genome. *Nature* 290 (5806): 457–65.

Anderson-Córdova, K. 1990. Hispaniola and Puerto Rico: Indian Acculturation and Heterogeneity, 1492–1550. Ph.D. diss., Yale University.

Andrews, R. M., I. Kubacka, P. F. Chinnery, R. N. Lightowlers, D. M. Turnbull, and N. Howell. 1999. Reanalysis and revision of the Cambridge Reference Sequence for human mitochondrial DNA. *Nature Genetics* 23 (2): 147.

Antón, S. 1989. Intentional cranial vault deformation and induced changes of the cranial base and face. *American Journal of Physical Anthropology* 79: 253–67.

Armstrong, D. 1980. Shellfish gatherers of St. Kitts: A study of Archaic subsistence and settlement patterns. In *Proceedings of the Eighth International Congress for the Study of the Pre-Columbian Cultures of the Lesser Antilles,* ed. S. Lewenstein, 152–67. Anthropological Research Papers 22. Tempe: Arizona State University.

Ascough, P., G. T. Cook, and A. J. Dugmore. 2005a. Methodological approaches to determining the marine radiocarbon reservoir effect. *Progress in Physical Geography* 29 (4): 532–47.

Ascough, P., G. T. Cook, A. J. Dugmore, M. Scott, and S. Freeman. 2005b. Influence of mollusk species on marine ΔR determinations. *Radiocarbon* 47 (3): 433–40.

Ayes, C. M. 1993. Angostura: Un campamento arcaico temprano del Valle Maunatabón, Bo. Florida Afuera, Barceloneta, Puerto Rico. Copies available at the Consejo para la Protección del Patrimonio Arqueológico Terrestre de Puerto Rico, San Juan.

Bailey, G. N. 1983. Concepts of time in Quaternary prehistory. *Annual Review of Anthropology* 12: 165–82.

Bamshad, M., T. Kivisild, W. S. Watkins, M. E. Dixon, C. E. Ricker, B. B. Rao, J. M. Naidu, et al. 2001. Genetic evidence on the origins of Indian caste populations. *Genome Research* 11 (6): 994–1004.

Bandelt, H. J., P. Forster, B. C. Sykes, and M. B. Richards. 1995. Mitochondrial portraits of human populations using median networks. *Genetics* 141 (2): 743–53.

Bandelt, H. J., V. Macaulay, and M. Richards. 2000. Median networks: Speedy construction and greedy reduction, one simulation, and two case studies from human mtDNA. *Molecular Phylogenetics and Evolution* 16 (1): 8–28.

Barbujani, F., and R. R. Sokal. 1990. Zones of sharp genetic change in Europe are also linguistic boundaries. *Proceedings of the National Academy of Sciences* 87: 1816–19.

Barker, H. 1970. Critical assessment of radiocarbon dating. *Philosophical Transactions of the Royal Society of London, Series A, Mathematical and Physical Sciences* 269 (1193): 37–45.

Batista, O., C. J. Kolman, and E. Bermingham. 1995. Mitochondrial DNA diversity in the Kuna Amerinds of Panamá. *Human Molecular Genetics* 4 (5): 921–29.

Batista Dos Santos, S. E., J. D. Rodrigues, A. K. C. Ribeiro-Dos-Santos, and M. A. Zago. 1999. Differential contribution of indigenous men and women to the formation of an urban population in the Amazon region as revealed by mtDNA and Y-DNA. *American Journal of Physical Anthropology* 109 (2): 175–80.

Bayliss, A., P. Marshall, and J. Sidell. 2004. A puzzling body from the river Thames in London. *Radiocarbon* 46 (1): 285–91.

Behar, D. M., M. G. Thomas, K. Skorecki, M. F. Hammer, E. Bulygina, D. Rosengarten, A. L. Jones, et al. 2003. Multiple origins of Ashkenazi Levites: Y chromosome evidence for both Near Eastern and European ancestries. *American Journal of Human Genetics* 73 (4): 768–79.

Bender, B. 1993. Introduction: Landscape-meeting and action. In *Landscape: Politics and perspectives,* ed. B. Bender, 1–17. Providence, R.I.: Berg.

Bentley, R. A. 2006. Strontium isotopes from the Earth to the archaeological skeleton: A review. *Journal of Archaeological Method and Theory* 13 (3): 135–87.

Bentley, R. A., T. D. Price, and E. Stephan. 2004. Determining the "local" $^{87}$Sr/$^{86}$Sr range for archaeological skeletons: A case study from Neolithic Europe. *Journal of Archaeological Science* 31: 365–75.

Berman, M. J., J. Febles, and P. L. Gnivecki. 2005. The organisation of Cuban archaeology: Context and brief history. In *Dialogues in Cuban archaeology*, ed. L. A. Curet, S. L. Dawdy, and G. La Rosa, 29–40. Tuscaloosa: University of Alabama Press.

Bernal, V., P. Novellino, P. N. Gonzalez, and S. I. Perez. 2007. Role of wild plant foods among Late Holocene hunter-gatherers from central and north Patagonia (South America): An approach from the dental evidence. *American Journal of Physical Anthropology* 133 (4): 1047–59.

Bert, F., A. Corella, M. Gené, et al. 2001. Major mitochondrial DNA haplotype heterogeneity in highland and lowland Amerindian populations from Bolivia. *Human Biology* 73: 1–16.

Binford, L. 1981. Behavioral archaeology and the "Pompeii Premise." *Journal of Anthropological Research* 37: 195–208.

Block, M. 1953. *The Historian's Craft*. Translated by P. Putnam. New York: Vintage Books.

Bolnick, D. A., and D. G. Smith. 2003. Unexpected patterns of mitochondrial DNA variation among Native Americans from the southeastern United States. *American Journal of Physical Anthropology* 122: 336–54.

Bonatto, S. L., and F. M. Salzano. 1997. Diversity and age of the four major mtDNA haplogroups, and their implications for the peopling of the New World. *American Journal of Human Genetics* 61: 1413–23.

Booden, M., R. Panhuysen, M. L. P. Hoogland, H. de Jong, G. Davies, and C. L. Hofman. 2006. Tracing human mobility in Guadeloupe with $^{87}$Sr. Paper presented at the symposium "New Methods and Techniques in the Study of Archaeological Materials from the Caribbean," 71st Annual Meeting of the Society for American Archaeology, San Juan, Puerto Rico.

Booden, M. A., R.G.A.M. Panhuysen, M. L. P. Hoogland, H. N. de Jong, G. R. Davies, and C. L. Hofman. 2008. Tracing human mobility with $^{87}$Sr/$^{86}$Sr at Anse à la Gourde, Guadeloupe. In *Crossing the borders: New methods and techniques in the study of archaeological materials from the Caribbean*, ed. C. L. Hofman, M. L. P. Hoogland, and A. L. van Gijn, 214–25. Tuscaloosa: University of Alabama Press.

Boomert, A. 2000. *Trinidad, Tobago and the Lower Orinoco Interaction Sphere: An archaeological/ethnohistorical study*. Alkmaar, The Netherlands: PlantijnCasparie Heerhugowaard B.V.

Boomert, A., and A. Bright. 2007. Island archaeology: In search of a new horizon. *Island Studies Journal* 2 (1): 3–26.

Bortolini, M. C., F. M. Salzano, M. G. Thomas, S. Stuart, S. P. K. Nasanen, C. H. D. Bau, M. H. Hutz, et al. 2003. Y-chromosome evidence for differing ancient demographic histories in the Americas. *American Journal of Human Genetics* 73 (3): 524–39.

Boudon, G., A. Le Friant, B. Villemant, and J.-P. Viode. 2005. Martinique. In Lindsay et al., *Volcanic hazard atlas of the Lesser Antilles,* 128–46.

Bower, B. 2000. Early New World settlers rise in East. *Science News* 157 (16): 244.

Breton, P. R. 1665. Dictionnaire Caraibe-Francoys, mesle de quantite de remarquest historiques pour l'eclaircissement de la langue. Auxerre: Gilles Bouquet.

Broecker, W. S., and E. A. Olson. 1961. Lamont radiocarbon measurements VIII: *Radiocarbon* 3: 176–204.

Brown, W. M., M. George Jr., and A. C. Wilson. 1979. Rapid evolution of animal mitochondrial DNA. *Proceedings of the National Academy of Sciences USA* 76 (4): 1967–71.

Bullen, R. P., and A. Bullen. 1972. *Archaeological investigations on St. Vincent and the Grenadines, West Indies.* William Bryant Foundation, American Studies, Report 8. Orlando, Fla.

Burney, D. A., and L. P. Burney. 1994. Holocene charcoal stratigraphy from Laguna Tortuguero, Puerto Rico, and the timing of human arrival on the island. *Journal of Archaeological Science* 21: 273–81.

Burns, J. A. 1996. Vertebrate paleontology and the alleged ice-free corridor: The meat of the matter. *Quaternary International* 32: 107–12.

Callaghan, R. 2001. Ceramic age seafaring and interaction potential in the Antilles: A computer simulation. *Current Anthropology* 42 (2): 308–13.

Callaghan, R. 2003a. Ceramic Age seafaring and interaction potential in the Antilles: A computer simulation. *Current Anthropology* 42: 308–13.

Callaghan, R. 2003b. Prehistoric trade between Ecuador and west Mexico: A computer simulation of coastal voyages. *Antiquity* 77: 796–804.

Callaghan, R. 2003c. Comments on the mainland origins of the preceramic cultures of the Greater Antilles. *Latin American Antiquity* 14 (3): 323–38.

Callaghan, R. 2007. Prehistoric settlement patterns on St. Vincent, West Indies. *Caribbean Journal of Science* 43: 11–22.

Callaghan, R., and S. Schwabe. 2001. Watercraft of the islands. In *Proceedings of the XVIII Congress of the International Association for Caribbean Archaeology, St. George, Grenada 1999,* ed. l'Association Internationale d'Archéologie de la Caraïbe, 231–42. Martinique: IACA.

Cann, R. L., M. Stoneking, and A. C. Wilson. 1987. Mitochondrial DNA and human evolution. *Nature* 325 (6099): 31–36.

Carini, S. P. 1991. *Compositional analysis of West Indian ceramics and their relevance to Puerto Rican prehistory.* Ph.D. diss., University of Connecticut. Ann Arbor, Mich.: University Microfilms.

Carvajal-Carmona, L. G., I. D. Soto, N. Pineda, D. Ortiz-Barrientos, C. Duque, J. Ospina-Duque, M. McCarthy, et al. 2000. Strong Amerind/white sex bias and a possible Sephardic contribution among the founders of a population in northwest Colombia. *American Journal of Human Genetics* 67 (5): 1287–95.

Carvalho-Silva, D. R., F. R. Santos, J. Rocha, and S. D. J. Pena. 2001. The phylogeography of Brazilian Y-chromosome lineages. *American Journal of Human Genetics* 68 (1): 281–86.

Chagnon, N. A. 1984. *Yanomamo: The fierce people.* New York: Holt, Rinehart and Winston.

Chandler, J. M. 2001. The Topper site: Beyond Clovis at Allendale. *Mammoth Trumpet* 16 (4): 10–15.

Chanlatte Baik, L. A. 1986. Cultura Ostionoide: Un desarrollo agroalfarero antilliano. *Homines* 10: 1–40.

Chanlatte Baik, L. A. 1990. Cultura Ostionoide: Un desarrollo agroalfarero antillano. In *Proceedings of the 11th Congress of the International Association for Caribbean Archaeology,* ed. G. Pantel, 295–311. San Juan: Fundación Arqueológica e Historica de Puerto Rico.

Chanlatte Baik, L. A., and I. Narganes. 1980. La Hueca, Vieques: Nuevo complejo cultural agroalfarero de la arqueología antillana. In *Proceedings of the Eighth International Congress for Caribbean Archaeology,* ed. S. Lewenstein, 501–23. Anthropological Research Papers 22. Tempe: Arizona State University.

Chanlatte Baik, L. A., and I. Narganes. 1983. *Vieques, Puerto Rico: Asiento de una nueva cultura aborigen antillana.* Santo Domingo, Dominican Republic: Impresora Corporán.

Chanlatte Baik, L. A., and R. N. Narganes Storde. 1986. *Processo desarrollode los primeros pobladores de Puerto Rico y las Antillas.* Santo Domingo, Dominican Republic: privately published.

Chikhi, L. R. Nichols, G. Barbujani, and M. A. Beaumont. 2002. Y genetic data support the Neolithic diffusion model. *Proceedings of the National Academy of Sciences* 99: 11008–13.

Clarke, J. 1989. *Atlantic pilot atlas.* Camden: International Marine.

Clayton, L. A., V. J. Knight Jr., and E. C. Moore. 1993. *The De Soto Chronicles,* vol. 1. Tuscaloosa: University of Alabama Press.

Cocilovo, J. A. 1973. Dimorfismo sexual y deformacion craneana artificial en Patagones del Chubut. *Acta II Simposio Internacional de Ciencias Morfologicas,* 633–42. Córdoba, Argentina.

Cocilovo, J. A. 1975. Estudio de dos factores que influencian la morfologia craneana en una coleccion andina: El sexo y la deformacion artificial. *Revista del Instituto Antropologico* 2: 197–212.

Cocilovo, J. A. 1978. Estudio de dos factores que influyen en la morfologia craneana de una coleccion patagonica: El sexo y la deformacion artificial. *Arquivos Anatomicos Antropologicos* 3: 113–41.

Cocilovo, J. A., and F. Rothhammer. 1990. Paleopopulation biology of the southern Andes—Craniofacial chronological and geographical differentiation. *Homo* 41: 16–31.

Cocilovo, J. A., and F. Rothhammer. 1996. Methodological approaches for the solution to ethnohistorical problems: Bioassay of kinship in prehistoric populations of Arica, Chile. *Homo* 47: 177–90.

Cocilovo, J. A., and F. Rothhammer. 1999. Morphological microevolution and extinction of kinship in prehistoric human settlements from the Azapata Valley, Chile. *Revista Chilena Historia Natural* 72: 213–18.

Cody, A. 1991. Distribution of exotic stone artifacts through the Lesser Antilles: Their implications for prehistoric interaction and exchange. In *Proceedings of the XIVth International Congress for Caribbean Archaeology,* ed. A. Cummins and P. King, 204–26. Bridgetown: Barbados Museum and Historical Society.

Conseil Général de la Martinique, Musée Départemental d'Archéologie Précolombienne et de Préhistoire. 2002 [1694]. *De Wilde ou les Sauvages Caribes Insulaires d'Amerique.* Martinique.

Cooper, J., R. Valcárcel Rojas, and P. Cruz Ramírez. 2006. Gente en los cayos: Los Buchillones y sus vínculos marítimos. *Caribe Arqueologico* 9: 66–75.

Coppa, A., A. Cucina, M. L. P. Hoogland, M. Lucci, F. Luna Calderón, R. Panhuysen, G. Tavarez, R. Valcárcel Rojas, and R. Vargiu. 2006. New evidence of two different migratory waves in the circum-Caribbean during the pre-Columbian period from the analysis of dental morphological traits. Paper presented at the symposium "New Methods and Techniques in the Study of Archaeological Materials from the Caribbean," 71st Annual Meeting of the Society for American Archaeology, San Juan, Puerto Rico.

Coppa, A., A. Cucina, M. Hoogland, M. Lucci, F. Luna Calderón, R. Panhuysen, G. Tavarez María, R. Valcárcel Rojas, and R. Vargiu. 2008. New evidence of two different migratory waves in the circum-Caribbean area during the pre-Columbian period from the analysis of dental morphological traits. In *Crossing the borders: New methods and techniques in the study of archaeology materials from the Caribbean,* ed. C. L. Hofman, M. L. P. Hoogland, and A. van Gijn, 195–213. Tuscaloosa: University of Alabama Press.

Coppa A., A. Cucina, M. Lucci, F. Luna Calderón, G. Tavarez, R. Vargiu. 2003. El poblamiento del area circum-caribeña: Afinidades biológicas y patron de migracíon desde el tercer milenio A.C. hasta la conquista. In *Proceedings of the XXth International Congress for Caribbean Archaeology,* vol. 2, ed. C. Tavárez Maria and M. A. García Arévalo, 493–506. Santo Domingo: Museo del Hombre Dominicano and Fundacíon Gracía Arevalo.

Coppa, A., A. Cucina, M. Lucci, A. Pellegrini, and R. Vargiu. 2002. The populations in the circum-Caribbean area from the 4th millennium BC to the conquest: The biological relationships according to possible migratory patterns. *American Journal of Physical Anthropology Supplement* 34: 57.

Crawford, M. H. 1986. Origin and maintenance of genetic variation on Black Carib populations of St. Vincent and Central America. In *Genetic variation and its maintenance, with particular reference to tropical populations,* ed. D. F. Roberts and G. De Stefano, 157–80. Cambridge: Cambridge University Press.

Crawford, M. H. 1998. *The origins of Native Americans.* Cambridge: Cambridge University Press.

Crawford, M. H., D. D. Dykes, K. Skradsky, et al. 1984. Blood group, serum protein, red cell enzyme polymorphisms, and admixture among the Black Caribs and Creoles of Central America and the Caribbean. In *Current developments in anthropological genetics,* vol. 3, *Black Caribs: A case study of biocultural adaptation,* ed. M. H. Crawford, 303–33. New York: Plenum Press.

Crawford, M. H., N. L. Gonzalez, M. S. Schanfield, et al. 1981. The Black Caribs (Garifuna) of Livingston, Guatemala: Genetic markers and admixture estimates. *Human Biology* 53: 87–103.

Crock, J. G. 2000. *Interisland interaction and development of chiefdoms in the Eastern Caribbean*. Ph.D. diss., University of Pittsburgh. Ann Arbor, Mich.: University Microfilms.

Crock, J. G., and J. B. Petersen. 1999. *A long and rich heritage: The Anguilla Archaeology Project, 1992–1998*. Prepared for the Anguilla Archaeological and Historical Society. Anguilla, British West Indies: The Valley.

Crock, J. G., and J. B. Petersen. 2004. Inter-island exchange, settlement hierarchy, and a Taíno-related chiefdom on the Anguilla Bank, Northern Lesser Antilles. In Delpuech and Hofman, *Late Ceramic Age societies in the eastern Caribbean*, 139–58.

Crock, J. G., J. B. Petersen, and N. Douglas. 1995. Preceramic Anguilla: A view from the Whitehead's Bluff site. In *Proceedings of the XVth International Congress of Caribbean Archaeology*, ed. R. Alegría and M. Rodríguez, 283–94. San Juan, P.R.: Centro de Estudios Avanzados de Puerto Rico y el Caribe.

Cruciani, F., P. Santolamazza, P. Shen, V. Macaulay, P. Moral, A. Olckers, D. Modiano, et al. 2000. A back migration from Asia to sub-Saharan Africa is supported by high-resolution analysis of human Y-chromosome haplotypes. *American Journal of Human Genetics* 70 (5): 1197–1214.

Curet, L. A. 2002. The chief is dead, long live . . . who? Descent and succession in the protohistoric chiefdoms of the Greater Antilles. *Ethnohistory* 49 (2): 259–80.

Curet, L. A. 2003. Issues on the diversity and emergence of middle-range societies of the ancient Caribbean: A critique. *Journal of Archaeological Research* 11 (1): 1–42.

Curet, L. A. 2004. Tibes: Un centro indígena temprano del centro-sur de Puerto Rico. *El Caribe Arqueológico* 6: 55–63.

Curet, L. A. 2005. *Caribbean paleodemography: Population, culture history, and sociopolitical processes in ancient Puerto Rico*. Tuscaloosa: University of Alabama Press.

Curet, L. A., S. L. Dawdy, and G. La Rosa Corzo, eds. 2005. *Dialogues in Cuban archaeology*. Tuscaloosa: University of Alabama Press.

Curet, L. A., L. A. Newsom, and S. DeFrance. 2006. Social and cultural change in the civic and ceremonial center of Tibes, Ponce, Puerto Rico. *Journal of Field Archaeology* 31: 23–39.

Curet, L. A., and J. R. Oliver. 1998. Mortuary practices, social development and ideology in pre-Columbian Puerto Rico. *Latin American Antiquity* 9 (3): 217–39.

Curet, L. A., J. M. Torres, and M. Rodriguez López. 2004. Political and social history of eastern Puerto Rico: The Ceramic Age. In Delpuech and Hofman, *Late Ceramic Age societies in the eastern Caribbean*, 59–86.

Dacal Moure, R. 2006. *Historiográfica arqueológica de Cuba*. Mexico City: Editorial Asesor Pedagógico.

Dacal Moure, R., and D. R. Watters. 2005. Three stages in the history of Cuban archaeology. In *Dialogues in Cuban archaeology*, ed. A. L. Curet, S. L. Dawdy, and G. La Rosa, 29–40. Tuscaloosa: University of Alabama Press.

Darroch, J. N., and J. E. Mosimann. 1985. Canonical and principal components of shape. *Biometrika* 72: 241–52.

Dávila-Dávila, O. 2003. *Arqueología de la Isla de la Mona*. San Juan: Instituto de Cultura Puertorriqueña.

Davis, D. D. 1974. Some notes concerning the Archaic occupation of Antigua. In *Proceedings of the Fifth International Congress for the Study of Pre-Columbian Cultures of the Lesser Antilles*, ed. R. P. Bullen, 65–71. Antigua: Antigua Archaeological Society.

Davis, D. D. 1988. Calibration of the Ceramic period chronology for Antigua, West Indies. *Southeastern Archaeology* 7 (1): 52–60.

Davis, D. D. 1993. Archaic blade production on Antigua, West Indies. *American Antiquity* 58 (4): 688–97.

Davis, D. D. 1996. Revolutionary archaeology in Cuba. *Journal of Archaeological Method and Theory* 3 (3): 159–88.

Davis, D. D. 2000. *Jolly Beach and the preceramic occupation of Antigua, West Indies*. Yale University Publications in Anthropology 84. New Haven, Conn.: Yale University Press.

Deetz, J. 1968. The inference of residence and descent rules from archaeological data. In *New perspectives in archaeology*, ed. S. R. Binford and L. R. Binford, 41–48. Chicago: Aldine.

Deevey, E. S., L. J. Gralenski, and V. Hoffren. 1959. Yale natural radiocarbon measurements IV. *Radiocarbon* 1 (1): 144–72.

Defense Mapping Agency Hydrographic/Topographic Center [DMAH/TC]. 1985. *Sailing directions for the Caribbean Sea*, vol. 1. Publication 147. Washington, D.C.

deFrance, S. D., and L. A. Newsom. 2005. The status of paleoethnobiological research on Puerto Rico and adjacent islands. In *Ancient Borinquen: Archaeology and ethnohistory of Native Puerto Rico*, ed. P. E. Siegel, 122–84. Tuscaloosa: University of Alabama Press.

Delpuech, A., and C. L. Hofman, eds. 2004. *Late Ceramic Age societies in the eastern Caribbean*. BAR International Series 1273. Oxford: Archaeopress.

de Mille, C. 2005. The preceramic occupation of Antigua, W.I. Ph.D. diss., University of Calgary.

Denaro, M., H. Blanc, M. J. Johnson, K. H. Chen, E. Wilmsen, L. L. Cavalli-Sforza, and D. C. Wallace. 1981. Ethnic variation in *Hpa*I endonuclease cleavage patterns of human mitochondrial DNA. *Proceedings of the National Academy of Sciences USA* 78 (9): 5768–72.

Departamento de Arqueología de Centro de Antropología. 2003. Atlas arqueológico de Cuba: Una estrategia científica para la investigación y la conservación del patrimonio histórico aborigen. *Catauro: Revista Cubana de Antropologia* 5 (8): 199–202.

Descantes, C., R. Speakman, M. D. Glascock, and D. V. Hill. 2005. Compositional analyses of ceramics from the Indian Creek site, Antigua: Preliminary results. Paper presented at the 21st International Congress for Caribbean Archaeology, Port of Spain, Trinidad.

de Waal, M. S. 2006. Pre-Columbian social organization and interaction interpreted through the study of settlement patterns: An archaeological case-study of the Pointe des Châteaux, La Désirade and Les Iles de la Petite Terre micro-region, Guadeloupe, F.W.I. Ph.D. diss., Leiden University.

Dillehay, T. D. 1997. *Monte Verde: A Late Pleistocene settlement in Chile,* vol. 2, *The archeological context and interpretation.* Washington, D.C.: Smithsonian Institution Press.

Dillehay, T. D. 1999. The late Pleistocene cultures of South America. *Evolutionary Anthropology* 7 (6): 206–16.

Dixon, E. J. 2001. Human colonization of the Americas: Timing, technology and process. *Quaternary Science Reviews* 20: 277–99.

Dixon, E. J. 2002. How and when did people first come to North America? *Athena Review* 3 (2): 23–27.

Donovan, Stephen K., R. K. Pickerill, R. W. Portell, T. A. Jackson, and D. A. T. Harper. 2003. The Miocene palaeobathymetry and palaeoenvironments of Carriacou, the Grenadines, Lesser Antilles. *Lethaia* 36: 255–72.

Drewett, P. L. 1995. Heywoods: Reconstructing a preceramic and late landscape on Barbados. In *Proceedings of the XV International Congress for Caribbean Archaeology,* ed. R. Alegría and M. Rodríguez, 273–82. San Juan, P.R.: Centro de Estudios Avansados de Puerto Rico y el Caribe y la Fundatión de las Humanidades y la Universidad del Turabo.

Drewett, P. L., ed. 2000. *Prehistoric settlements in the Caribbean: Fieldwork in Barbados, Tortola, and the Cayman Islands.* London: Archetype Publications.

Easton, R. D., D. A. Merriwether, D. E. Crews, et al. 1996. mtDNA variation in the Yanomami: Evidence for additional New World founding lineages. *American Journal of Human Genetics* 59: 213–25.

Ember, M. 1974. The conditions that may favor avunculocal residence. *Behavior Science Review* 9: 203–9.

Ember, M., and C. R. Ember. 1971. The conditions favoring matrilocal versus patrilocal residence. *American Anthropologist,* n.s., 73: 571–94.

Evans, J. A., C. A. Chenery, and A. P. Fitzpatrick. 2006. Bronze Age childhood migration of individuals near Stonehenge, revealed by strontium and oxygen isotope tooth enamel analysis. *Archaeometry* 48 (2): 309–21.

Ezzo, J. A., C. M. Johnson, and T. D. Price. 1997. Analytical perspectives on prehistoric migration: A case study from east-central Arizona. *Journal of Archaeological Science* 24: 447–66.

Ezzo, J. A., and T. D. Price. 2002. Migration, regional reorganization, and spatial group composition at Grasshopper Pueblo, Arizona. *Journal of Archaeological Science* 29: 499–520.

Fabian, J. 1983. *Time and the Other: How anthropology makes its object.* New York: Columbia University Press.

Fabra, M., A. G. Laguens, and D. A. Demarchi. 2007. Human colonization of the central territory of Argentina: Design matrix models and craniometric evidence. *American Journal of Physical Anthropology* 133 (4): 1060–66.

Fagundes, N. J. R., R. Kanitz, and S. L. Bonatto. 2008. A reevaluation of the Native American mtDNA genome diversity and its bearing on the models of early colonization of Beringia. *PLoS ONE* 3 (9): e3157.

Falsetti, A. B., W. L. Jungers, and T. M. Cole. 1993. Morphometrics of the callitrichid fore-limb: A case study in size and shape. *International Journal of Primatology* 14: 551–72.

Febles, J. 1982. *Estudio tipológico y tecnológico del material de piedra tallada del sitio arqueológico Canímar I, Matanzas, Cuba.* Havana: Editora de la Academia de Ciencias de Cuba.

Febles Duenas, J., J. M. Guarch Delmonte, A. Rives, R. Sánchez, and M. Monteagudo. 1987. *Censo arqueológico de Cuba por tratamiento computarizado.* Folleto (brochure). Havana: Departamento de Arqueología, Centro de Antropología, Academia de Ciencias de Cuba.

Febles Duenas, J., and A. R. Martínez. 1995. Informacion censal arqueologica de Cuba. In *Taíno: Arqueologia de Cuba.* CD-ROM. Colima: Centro de Antropología y CEDISAC.

Fedje, D. W., and T. Christensen. 1999. Modeling paleoshorelines and locating early Holocene coastal sites in Haida Gwai. *American Antiquity* 64 (4): 635–52.

Fedje, D. W., Q. Mackie, E. J. Dixon, et al. 2004. Late Wisconsin environments and archaeological visibility on the northern Northwest Coast. In *Entering America: Northeast Asia and Beringia before the Last Glacial Maximum,* ed. D. Madsen, 97–138. Salt Lake City: University of Utah Press.

Fedje, D. W., R. J. Wigen, Q. Mackie, et al. 2001. Preliminary results from investigations at Kilgii Gwaay: An early Holocene archaeological site on Ellen Island, Haida Gwaii, British Columbia. *Canadian Journal of Archaeology* 25 (1–2): 98–120.

Feliciano-Vélez, A. 2006. Genetic prints of Amerindian female migrations through the Caribbean revealed by control sequences from Dominican haplogroup A mtDNAs. M.S. thesis, University of Puerto Rico, Mayagüez.

Fernández-Méndez, E. 1970. *Historia cultural de Puerto Rico.* San Juan: Ediciones El Cemí.

Fewkes, J. W. 1907. The aborigines of Porto Rico and neighbouring islands. In *Twenty-Fifth annual report of the U.S. Bureau of Ethnology to the secretary of the Smithsonian Institution, 1903–04.* New York: Johnson Reprint Corporation, 1970.

Fewkes, J. W. 1914. Relations of aboriginal culture and environment in the Lesser Antilles. *Bulletin of the American Geographical Society* 46 (9): 667–68.

Fitzpatrick, S. M. 2006. A critical approach to [14]C dating in the Caribbean: Using chronometric hygiene to evaluate chronological control and prehistoric settlement. *Latin American Antiquity* 17 (4): 389–418.

Fitzpatrick, S. M., M. Kappers, and Q. Kaye. 2006. Coastal erosion and site destruction on Carriacou, West Indies. *Journal of Field Archaeology* 31 (3): 251–62.

Fitzpatrick, S. M., M. Kappers, Q. Kaye, C. M. Giovas, M. J. LeFebvre, M. H. Harris, S. Burnett, J. A. Pavia, K. M. Marsaglia, and J. Feathers. 2010. Precolumbian settlements on Carriacou, West Indies. *Journal of Field Archaeology* 34: 247–66.

Fitzpatrick, S. M., Q. Kaye, J. Feathers, J. A. Pavia, K. M. Marsaglia. 2009. Evidence for inter-island transport of heirlooms: Luminescence dating and petrographic analysis of ceramic inhaling bowls from Carriacou, West Indies. *Journal of Archaeological Science* 36: 596–606.

Fitzpatrick, S. M., Q. Kaye, and M. Kappers. 2004. A radiocarbon sequence for the Sabazan site, Carriacou, Grenadines. *Journal of Caribbean Archaeology* 5: 1–11.

Fitzpatrick, S. M., and W. F. Keegan. 2007. Human impacts and adaptations in the Caribbean islands: An historical ecology approach. *Earth and Environmental Science Transactions of the Royal Society of Edinburgh* 98: 29–45.

Fix, A. G. 2005. Rapid deployment of the five founding Amerind mtDNA haplogroups via coastal and riverine colonization. *American Journal of Physical Anthropology* 128 (2): 430–36.

Fladmark, K. R. 1979. Routes: Alternative migration corridors for early man in North America. *American Antiquity* 44: 55–69.

Foreman, G. 1953. *Indian removal: The emigration of the Five Civilized Tribes of Indians.* Norman: University of Oklahoma Press.

Forster, P., R. Harding, A. Torroni, and H. J. Bandelt. 1996. Origin and evolution of Native American mtDNA variation: A reappraisal. *American Journal of Human Genetics* 59 (4): 935–45.

Forster, P., and S. Matsumura. 2005. Did early humans go north or south? *Science* 308 (5724): 965–66.

Forster, P., A. Röhl, P. Lünnemann, C. Brinkmann, T. Zerjal, C. Tyler-Smith, and B. Brinkmann. 2000. A short tandem repeat-based phylogeny for the human Y chromosome. *American Journal of Human Genetics* 67 (1): 182–96.

Fuselli, S., E. Tarazona-Santos, I. Dupanloup, A. Soto, D. Luiselli, and D. Pettener D. 2003. Mitochondrial DNA diversity in South America and the genetic history of Andean highlanders. *Molecular Biology and Evolution* 20 (10): 1682–91.

Galloway, P., ed. 1997. *The Hernando de Soto Expedition.* Lincoln: University of Nebraska Press.

Garrow, P. H., C. H. McNutt Jr., G. G. Weaver, and J. R. Oliver. 1995. La Iglesia de Maraguez (PO-39): Investigations of a local ceremonial center in the Cerrillos River Valley, Ponce, Puerto Rico. Contract report submitted to the U.S. Army Corps of Engineers, Jacksonville. On file at the State Historic Preservation Office, San Juan, Puerto Rico.

Giles, R. E., H. Blanc, H. M. Cann, and D. C. Wallace. 1980. Maternal inheritance of human mitochondrial DNA. *Proceedings of the National Academy of Sciences USA* 77 (11): 6715–19.

Godo, P. 1997. El problema del protoagrícola de Cuba: Discusión y perspectivas. *Caribe Arqueologico* 2: 19–30.

Godo Torres, P. P. 1994. Industrias de la concha y de la piedra no lascada del sitio arqueológico mesolítico Victoria I, provincia de Camaguey. In *Estudios arqueologicos,* ed. J. Febles, L. Dominguez, F. Ortega, G. La Rosa, A. Martinez, and A. Rives, 140–65. Havana: Editorial Academia.

González Colón, J. 1984. Tibes: Un centro ceremonial indígena. Master's thesis, Centro de Estudios Avanzados de Puerto Rico y el Caribe, San Juan.

González-José, R., W. A. Neves, M. M. Lahr, S. Gonzalez, H. M. Pucciarelli, and G. Correal. 2005. Late Pleistocene/Holocene craniofacial morphology in Mesoamerican

Paleoindians: Implications for the peopling of the New World. *American Journal of Physical Anthropology* 128: 772–80.

González-Oliver, A., L. Marquez-Morfin, J. C. Jimenez, et al. 2001. Founding Amerindian mitochondrial DNA lineages in ancient Maya from Xcaret, Quintana Roo. *American Journal of Physical Anthropology* 116 (3): 230–35.

Goodenough, W. H. 1955. Residence rules. *Southwestern Journal of Anthropology* 12: 22–37.

Goodfriend, G. A., and D. G. Hood. 1983. Carbon isotope analysis of land snail shells: Implications for carbon sources and radiocarbon dating. *Radiocarbon* 25 (3): 810–30.

Goodyear, A., K. Steffy, K. B. Sweeney, et al. 2005. Clovis in Allendale: The Topper and Big Pine Tree sites. Paper presented at the Clovis in the Southeast Conference, Columbia, S.C.

Green, L. D., J. N. Derr, and A. Knight. 2000. mtDNA affinities of the peoples of north-central Mexico. *American Journal of Human Genetics* 66: 989–98.

Greenberg, B. D., J. E. Newbold, and A. Sugino. 1983. Intraspecific nucleotide sequence variability surrounding the origin of replication in human mitochondrial DNA. *Gene* 21 (1–2): 33–49.

Greenberg, J. H. 1987. *Language in the Americas*. Stanford: Stanford University Press.

Greenberg, J. H., and M. Ruhlen. 1992. Linguistic origins of Native Americans. *Scientific American* 267: 94–99.

Greenberg, J. H., C. G. Turner II, and S. L. Zegura. 1986. The settlement of the Americas: A comparison of the linguistic, dental, and genetic evidence. *Current Anthropology* 27: 477–97.

Guarch Delmonte, J. M. 1978. *El Taino de Cuba: Ensayo de reconstrucción etno-histórica*. Havana: Academia de Ciencias de Cuba.

Guarch Delmonte, J. M. 1990. *Estructura para las comunidades aborigenes de Cuba*. Holguín: Ediciones Holguín.

Guarch Delmonte, J. M., E. E. Rey Betancourt, and J. Febles Duenas. 1995. Historía aborígen de Cuba. In *Taíno: Arqueologia de Cuba*. CD-ROM. Colima: Centro de Antropología y CEDISAC.

Hackenberger, S. 1991. Archaeological test excavation of Buccament Valley Rockshelter, St. Vincent: Preceramic stone tools in the Windward Islands, and the early peopling of the eastern Caribbean. In *Proceedings of the XIII International Congress for Caribbean Archaeology*, ed. A. Pantel T., I. Vargas A., and M. Sanoja O., 86–91. Curaçao: IACA.

Hage, P. 1998. Was proto-Oceanic society matrilineal? *Journal of the Polynesian Society* 107: 365–79.

Hage, P. 1999. Reconstructing ancestral Oceanic society. *Asian Perspectives* 38: 200–228.

Hage, P., and F. Harary. 1983. *Structural models in anthropology*. Cambridge: Cambridge University Press.

Hage, P., and F. Harary. 1996. *Island networks: Communication, kinship, and classification structures in Oceania*. Cambridge: Cambridge University Press.

Hage, P., and J. Marck. 2003. Matrilineality and the Melanesian origin of Polynesian Y chromosomes. *Current Anthropology* 44 (supplement): S121–S127.

Hammer, M. F., T. M. Karafet, A. J. Redd, H. Jarjanazi, S. Santachiara-Benerecetti, H. Soodyall, and S. L. Zegura. 2001. Hierarchical patterns of global human Y-chromosome diversity. *Molecular Biology and Evolution* 18 (7): 1189–1203.

Hammer, M. F., A. J. Redd, E. T. Wood, M. R. Bonner, H. Jarjanazi, T. Karafet, S. Santachiara-Benerecetti, et al. 2000. Jewish and Middle-Eastern non-Jewish populations share a common pool of Y-chromosome biallelic haplotypes. *Proceedings of the National Academy of Sciences* 97 (12): 6769–74.

Hammer, M. F., A. B. Spurdle, T. Karafet, M. R. Bonner, E. T. Wood, A. Novelletto, P. Malaspina, et al. 1997. The geographic distribution of human Y chromosome variation. *Genetics* 145 (3): 787–805.

Harris, M. H. 2005. Introduction to the pottery of Carriacou. Paper presented at the 21st Congress of the International Association of Caribbean Archaeology, Trinidad, July 26.

Haviser, J. 1991. *The first Bonaireans.* Reports of the Archaeological-Anthropological Institute of the Netherlands Antilles 10. Curaçao.

Haviser, J. 1997. Settlement strategies in the Early Ceramic Age. In *The indigenous people of the Caribbean,* ed. S. M. Wilson, 59–69. Gainesville: University Press of Florida.

Haydenblit, R. 1996. Dental variation among four Prehispanic Mexican populations. *American Journal of Physical Anthropology* 100: 225–46.

Haynes, C. V., Jr. 1993. Clovis-Folsom geochronology and climatic change. In *From Kostenki to Clovis,* ed. O. Soffer and N. D. Praslov, 219–26. New York: Plenum Press.

Heaton, T. H., and F. Grady. 2003. The Late Wisconsin vertebrate history of Prince of Wales Island, southeast Alaska. In *Vertebrate paleontology of Late Cenozoic cave deposits in North America,* ed. B. W. Schubert, J. I. Mead, and R. W. Graham, 17–53. Bloomington: Indiana University Press.

Hernández Oliva, C. A., and R. Arrazcaeta Delgado. 2004. Prehistoria de Cuba: Una propuesta de análisis teórica y metodológico. *Caribe Arqueologico* 8: 64–73.

Hetherington, R., and R. G. B. Reid. 2003. Malacological insights into the marine ecology and changing climate of the late Pleistocene–early Holocene Queen Charlotte Islands archipelago, western Canada, and implications for early peoples. *Canadian Journal of Zoology* 81: 626–61.

Hey, J. 2005. On the number of New World founders: A population genetic portrait of the peopling of the Americas. *PLoS Biology* 3 (6): e193.

Higham, T. 2005. Radiocarbon Web-info: The method. http://www.c14dating.com/ (last accessed May 8, 2007).

Hill, J. N. 1970. *Broken K Pueblo.* Anthropology Papers of the University of Arizona 18. Tucson: University of Arizona Press.

Hill, J. N., ed. 1977. *The explanation of prehistoric change.* Albuquerque: University of New Mexico Press.

Hill, J. N., and J. Gunn, eds. 1977. *The individual in prehistory: Studies of variability in style in prehistoric technologies.* New York: Academic Press.

Hincks, T., S. Sparks, P. Dunkley, and P. Cole. 2005. Montserrat. In Lindsay et al., *Volcanic hazard atlas of the Lesser Antilles*, 148–68.

Hodell, D. A., R. L. Quinn, M. Brenner, and G. Kamenov. 2004. Spatial variation of strontium isotopes ($^{87}Sr/^{86}Sr$) in the Maya region: A tool for tracking ancient human migration. *Journal of Archaeological Science* 31: 585–601.

Hofman, C. L. 1993. In search of the native population of pre-Columbian Saba (400–1450 AD), part one: Pottery styles and their interpretation. Ph.D. diss., Leiden University.

Hofman, C. L., A. Boomert, A. J. Bright, M. L. P. Hoogland, S. Knippenberg, and A. V. Samson. 2006b. Ties with the homeland: Archipelagic interaction and the enduring role of the South American mainland in the pre-Columbian Caribbean. Paper presented at the opening session, "Islands in the Stream: Interisland and Continental Interaction in the Caribbean," 71st Annual Meeting of the Society for American Archaeology, San Juan, Puerto Rico.

Hofman, C. L., A. J. Bright, and M. L. P. Hoogland. 2006a. Archipelagic resource procurement and mobility in the northern Lesser Antilles: The view from a 3000-year-old tropical forest campsite on Saba. *Journal of Island and Coastal Archaeology* 1:145–64.

Hofman, C. L., A. Bright, M. L. P. Hoogland, and W. F. Keegan. 2008. Attractive ideas, desirable goods: Examining the Late Ceramic Age relationships between Greater and Lesser Antillean societies. *Journal of Island and Coastal Archaeology* 3: 17–34.

Hofman, C. L., A. Delpuech, and M. L. P. Hoogland. 1999. Excavations at the site of Anse à la Gourde, Guadeloupe: Stratigraphy, ceramic chronology and structures. In *Proceedings of the XVIIIth International Congress for Caribbean Archaeology*, vol. 2, 162–72. Guadeloupe: International Association for Caribbean Archaeology.

Hofman, C. L., and M. L. P. Hoogland. 2003. Plum Piece: Evidence for Archaic seasonal occupation on Saba, northern Lesser Antilles, around 3300 BP. *Journal of Caribbean Archaeology* 4: 1–16.

Hofman, C. L., and M. L. P. Hoogland. 2004. Social dynamics and change in the northern Lesser Antilles. In Delpuech and Hofman, *Late Ceramic Age societies in the eastern Caribbean*, 33–44.

Hofman, C. L., M. L. P. Hoogland, and A. Delpuech. 2001. Social organization at a Troumassoid settlement: The case of Anse à la Gourde, Guadeloupe. In *Proceedings of the XIXth International Congress for Caribbean Archaeology* [Aruba 2001], vol. 2, ed. L. Alofs and R.A.C.F. Dijkhoff, 124–31. Museo Arqueologico 9. Oranjestad, Aruba: Museo Arqueologico.

Hofman, C. L., A. J. D. Isendoorn, M. Booden, and L. Jacobs. 2006c. In tuneful threefold: Combining conventional archaeological methods, geochemical analysis and ethnoarchaeological research in the study of pre-colonial pottery of the Caribbean. Paper presented at the symposium "New Methods and Techniques in the Study of Archaeological Materials from the Caribbean," 71st Annual Meeting of the Society for American Archaeology, San Juan, Puerto Rico.

Hoogland, M. L. P., T. Romon, and P. Brasselet. 1999. Troumassoid burial practices at the site of Anse à la Gourde, Guadeloupe. In *Proceedings of the XVIIIth International Congress for Caribbean Archaeology* [Grenada 1999], vol. 2, 173–78. Guadeloupe: International Association for Caribbean Archaeology.

Howells, W. W. 1973. *Cranial variation in man: A study by multivariate analysis of patterns of difference among recent human populations.* Papers of the Peabody Museum of Archaeology and Ethnology 67. Cambridge, Mass.: Harvard University.

Hrdlička, A. 1920. Shovel-shaped teeth. *American Journal of Physical Anthropology* 3: 429–65.

Hudson, C. 1997. *Knights of Spain, warriors of the sun: Hernando de Soto and the South's ancient chiefdoms.* Athens: University of Georgia Press.

Hughen, K. A., M. G. L. Baillie, E. Bard, J. W. Beck, C. J. H. Bertrand, P. G. Blackwell, C. E. Buck, et al. 2004. Marine04 marine radiocarbon age calibration, 0–26 cal kyr BP. *Radiocarbon* 46 (3): 1059–86.

Hughen, K. A., J. T. Overpeck, L. C. Peterson, and R. F. Anderson. 1996. The nature of varved sedimentation in the Cariaco Basin, Venezuela, and its palaeoclimatic significance. In *Palaeoclimatology and palaeoceanography from laminated sediments, A.E.,* ed. S. Kemp, 171–83. Geological Society Special Publication 116. Bath, U.K.

Huoponen, K., A. Torroni, P. R. Wickman, et al. 1997. Mitochondrial and Y chromosome–specific polymorphisms in the Seminole tribe of Florida. *European Journal of Human Genetics* 5: 25–34.

Hutchison, C. A., J. E. Newbold, S. S. Potter, and M. H. Edgell. 1974. Maternal inheritance of mammalian mitochondrial DNA. *Nature* 251 (5475): 536–38.

Ingold, T. 2000. *The perception of the environment: Essays on livelihood, dwelling and skill.* London: Routledge.

Irwin, G. 1992. *The prehistoric exploration and colonization of the Pacific.* Cambridge: Cambridge University Press.

Jantz, R. L., and D. Owsley. 2001. Variation among early North American crania. *American Journal of Physical Anthropology* 114: 146–55.

Jardines Macias, J., and J. Calvera Roses. 1999. Estructuras de viviendas aborígenes en Los Buchillones. *Caribe Arqueologico* 3: 44–52.

Johnson, M. J., D. C. Wallace, S. D. Ferris, M. C. Rattazzi, and L. L. Cavalli-Sforza. 1983. Radiation of human mitochondria DNA types analyzed by restriction endonuclease cleavage patterns. *Journal of Molecular Evolution* 19 (3–4): 255–71.

Josenhans, H. W., D. W. Fedje, R. Pienitz, et al. 1997. Early humans and rapidly changing Holocene sea levels in the Queen Charlotte Islands–Hecate Strait, British Columbia, Canada. *Science* 277: 71–74.

Jouravleva, I., and N. González. 2000. Las varitationes climáticas y la reutilización del espacio habitacional a través de la alfarería Aborigen. *Caribe Arqueologico* 4: 35–39.

Jull, A. J. T., M. Iturralde-Vinent, J. M. O'Malley, R. D. E. MacPhee, H. G. McDonald, P. S. Martin, J. Moody, and A. Rincón. 2004. Radiocarbon dating of extinct fauna in the Americas recovered from tar pits. *Nuclear Instruments and Methods in Physics Research B* 223–24: 668–71.

Kaestle, F. A., and D. G. Smith. 2001. Ancient mitochondrial DNA evidence for prehistoric population movement: The Numic expansion. *American Journal of Physical Anthropology* 115 (1): 1–12.

Kaye, Q. P. 2003. A field survey of the island of Carriacou, West Indies, March 2003. *Papers from the Institute of Archaeology* 14: 129–35.

Kaye, Q. P., S. M. Fitzpatrick, J. A. Carstensen, K. M. Marsaglia, and J. Feathers. 2009. Evidence for inter-island transport of heirlooms? Petrographic analysis and luminescence dating of ceramic inhaling bowls from Carriacou, West Indies. Paper presented at the XXI International Congress for Caribbean Archaeology, Kingston, Jamaica.

Kaye, Q. P., M. Kappers, and S. M. Fitzpatrick. 2004. An archaeological survey of Carriacou, West Indies. In *Proceedings of the XX International Congress for Caribbean Archaeology*, 391–98. Santo Domingo, Dominican Republic.

Keegan, W. F. 1985. *Dynamic horticulturalists: Population expansion in the prehistoric Bahamas*. Ph.D. diss., University of California, Los Angeles. Ann Arbor, Mich.: University Microfilms.

Keegan, W. F. 1989. Creating the Guanahatabey (Ciboney): The modern genesis of an extinct culture. *Antiquity* 63 (239): 373–79.

Keegan, W. F. 1992. *The people who discovered Columbus: The prehistory of the Bahamas*. Gainesville: University Press of Florida.

Keegan, W. F. 1994. West Indian archaeology 1: Overview and foragers. *Journal of Archaeological Research* 2 (3): 255–84.

Keegan, W. F. 1995. Modeling dispersal in the prehistoric West Indies. *World Archaeology* 26: 400–420.

Keegan, W. F. 1996. West Indian archaeology 2: After Columbus. *Journal of Archaeological Research* 2 (3): 265–94.

Keegan, W. F. 2000. West Indian archaeology 3: Ceramic Age. *Journal of Archaeological Research* 8 (2): 135–67.

Keegan, W. F. 2004. Islands of chaos. In Delpuech and Hofman, *Late Ceramic Age societies in the eastern Caribbean*, 33–44.

Keegan, W. F. 2006a. Archaic influences in the origins and development of Taíno societies. *Caribbean Journal of Science* 42: 1–13.

Keegan, W. F. 2006b. All in the family: Descent and succession in the protohistoric chiefdoms of the Greater Antilles—A comment on Curet. *Ethnohistory* 53: 383–92.

Keegan, W. F. 2007. *Taíno Indian myth and practice: The arrival of the stranger king*. Gainesville: University Press of Florida.

Keegan, W. F. 2009. Central plaza burials in Saladoid Puerto Rico: An alternative perspective. *Latin American Antiquity* 20:375–85.

Keegan, W. F. 2010. Demographic imperatives for island colonists. In *The global origins and development of seafaring*, ed. A. J. Anderson, H. H. Barrett, and K. V. Boyle, 171–78. Cambridge, UK: McDonald Institute for Archaeological Research.

Keegan, W. F., and J. Diamond. 1987. Colonization of islands by humans: A biogeographical perspective. In *Advances in archaeological method and theory*, ed. M. Schiffer, 49–92. New York: Academic Press.

Keegan, W. F., and M. D. Maclachlan. 1989. The evolution of avunculocal chiefdoms: A reconstruction of Taino kinship and politics. *American Anthropologist* 91: 613–30.

Keegan, W. F., M. D. Maclachlan, and B. Byrne. 1998. Social foundations of the Taino caciques. In *Chiefdoms and chieftaincy in the Americas*, ed. E. Redmond, 217–44. Gainesville: University Press of Florida.

Kemp, B. M., A. Resendez, J. A. Román-Berrelleza, et al. 2005. An analysis of ancient

Aztec mtDNA from Tlatelolco: Pre-Columbian relations and the spread of Uto-Aztecan. In *Biomolecular archaeology: Genetic approaches to the past*, ed. D. Reed, 22–46. Center for Archeological Investigations, Occasional Paper 32. Carbondale: Southern Illinois University Press.

Kenny, J. S. 1989. Hermatyic scleractinian corals of Trinidad. *Studies in the Fauna of Curacao and Other Caribbean Islands* 123: 83–100.

Keyeux, G., C. Rodas, N. Gelvez, et al. 2002. Possible migration routes into South America deduced from mitochondrial DNA studies in Colombian Amerindian populations. *Human Biology* 74 (2): 211–33.

Kirch, P. V. 2000. *On the road of the winds*. Berkeley: University of California Press.

Kitchen, A., M. M. Miyamoto, and C. J. Mulligan. 2008. A three-stage colonization model for the peopling of the Americas. *PLoS ONE* 3 (2): e1596.

Knapp, A. B. 1992. Archaeology and annales: Time, space and change. In *Archaeology, annales and ethnohistory*, ed. A. B. Knapp, 1–21. Cambridge: Cambridge University Press.

Knippenberg, S. 1995. Norman Estate and Anse des Peres: Two Precolumbian sites on St. Martin. Master's thesis, Leiden University.

Knippenberg, S. 2006. Stone artefact production and exchange among the northern Lesser Antilles. Ph.D. diss., Leiden University.

Knudson, K. J., K. C. Nystrom, T. A. Tung, T. D. Price, and P. D. Fullagar. 2005. The origin of the Juch'uypampa cave mummies: Strontium isotope analysis of archaeological human remains from Bolivia. *Journal of Archaeological Science* 32: 903–13.

Knudson, K. J., T. D. Price, J. E. Buikstra, and D. E. Blom. 2004. The use of strontium isotope analysis to investigate Tiwanaku migration and mortuary ritual in Bolivia and Peru. *Archaeometry* 46: 5–18.

Koch, P. L., and A. D. Barnosky. 2006. Late quaternary extinctions: State of the debate. *Annual Review of Ecology Evolution and Systematics* 37: 215–50.

Kolman, C. J., and E. Bermingham. 1997. Mitochondrial and nuclear DNA diversity in the Choco and Chibcha Amerinds of Panama. *Genetics* 147: 289–302.

Kolman, C. J., E. Bermingham, R. Cooke, R. H. Ward, T. D. Arias, and F. Guionneau-Sinclair. 1995. Reduced mtDNA diversity in the Ngöbé Amerinds of Panamá. *Genetics* 140 (1): 275–283.

Komorowski, J.-C., G. Boudon, M. Semet, F. Beauducel, C. Antenor-Habazac, S. Bazin, and G. Hammouya. 2005. Guadeloupe. In Lindsay et al., *Volcanic hazard atlas of the Lesser Antilles*, 68–106.

Kozlowski, J. K. 1974. *Preceramic cultures in the Caribbean*. Kraków: Prace Archeologiczne.

Krause, R. A. 1989. Coffee, sugar and baked clay: From prehistory to history in Puerto Rico's Cerrillos River valley. Report submitted to the U.S. Army Corps of Engineers, Jacksonville District. Copies available from the Puerto Rico State Historic Preservation Office, San Juan.

Lahr, M. M. 1995. Patterns of modern human diversification: Implications for Amerindian origins. *Yearbook of Physical Anthropology* 38: 163–98.

Lalueza-Fox, C., M. T. P. Gilbert, A. Martínez-Fuentes, F. Calafell, and J. Betranpetit.

2003. Mitochondrial DNA from pre-Columbian Ciboneys from Cuba and the pre-historic colonization of the Caribbean. *American Journal of Physical Anthropology* 121: 97–108.

Lalueza-Fox, C., F. Luna Calderon, F. Calafell, B. Morera, and J. Bertranpetit. 2001. mtDNA from extinct Tainos and the peopling of the Caribbean. *Annals of Human Genetics* 65 (2): 137–51.

Lammers-Keijsers, Y. M. J. 2007. *Tracing traces from present to past: A functional analysis of pre-Columbian shell and stone artefacts from Anse à la Gourde and Morel, Guadeloupe, FWI.* Leiden: Leiden University Press.

La Rosa Corzo, G. 2003. La ciencia arqueológica en Cuba: Retos y perspectivas en los umbrales del siglo XXI. *Catauro: Revista Cubana de Antropologia* 5 (8): 36–46.

LeFebvre, M. J. 2007. Zooarchaeological analysis of prehistoric vertebrate exploitation at the Grand Bay Site, Carriacou, West Indies. *Coral Reefs* 26: 941–44.

Lell, J. T., R. I. Sukernik, Y. B. Starikovskaya, B. Su, L. Jin, T. G. Schurr, P. A. Underhill, and D. C. Wallace. 2002. The dual origin and Siberian affinities of Native American Y chromosomes. *American Journal of Human Genetics* 70 (1): 192–206.

Lewis, C. M., R. Y. Tito, B. Lizárraga, and A. C. Stone. 2005. Land, language, and loci: mtDNA in Native Americans and the genetic history of Peru. *American Journal of Physical Anthropology* 127 (3): 351–60.

Li, W-H. 1997. *Molecular evolution.* Sunderland, Mass.: Sinauer Associates.

Lindsay, J. 2005. St. Lucia. In Lindsay et al., *Volcanic hazard atlas of the Lesser Antilles,* 219–39.

Lindsay, J., R. Robertson, J. Shepard, and S. Ali, eds. 2005a. *Volcanic hazard atlas of the Lesser Antilles.* Trinidad and Tobago: Seismic Research Unit of the University of the West Indies.

Lindsay, J., and J. Shepard. 2005. Kick 'em Jenny and Ile de Caille. In Lindsay et al., *Volcanic hazard atlas of the Lesser Antilles,* 108–26.

Lindsay, J., A. Smith, M. Roobol, and M. Stasiuk. 2005b. Dominica. In Lindsay et al., *Volcanic hazard atlas of the Lesser Antilles,* 2–47.

Liston, J. 2005. An assessment of radiocarbon dates from Palau, Micronesia. *Radiocarbon* 47: 295–354.

Longacre, W. A. 1970. *Archaeology as anthropology: A case study.* Anthropology Papers of the University of Arizona 17. Tucson: University of Arizona Press.

Lorenz, J. G., and D. G. Smith. 1997. Distribution of sequence variation in the mtDNA control region of Native North Americans. *Human Biology* 69: 749–76.

Luis, J. R., D. J. Rowold, M. Regueiro, B. Caeiro, C. Cinnioğlu, C. Roseman, P. A. Underhill, L. L. Cavalli-Sforza, and R. J. Herrera. 2004. The Levant versus the Horn of Africa: Evidence for bidirectional corridors of human migrations. *American Journal of Human Genetics* 74 (3): 532–44.

Lumholtz, C., and A. Hrdlička. 1898. Marked human bones from a prehistoric Tarasco Indian burial place in the State of Michoacan, Mexico. *Bulletin of the AMNH.* 10: 61–90.

Lundberg, E. R. 1985. Settlement pattern analysis for south central Puerto Rico. In *Archaeological data recovery at El Bronce, Puerto Rico—final report, phase 2,* ed. L. S.

Robinson, 1–23 (Appendix L). Archaeological Services, Inc., contract report submitted to the U.S. Army Corps of Engineers, Jacksonville. On file at the State Historic Preservation Office, San Juan, Puerto Rico.

Lundberg, E. R. 1989. *Preceramic procurement patterns at Krum Bay, Virgin Islands*. Ph.D. diss., University of Illinois, Urbana-Champagne. Ann Arbor, Mich.: University Microfilms.

Maat, G. J. R., R.G.A.M. Panhuysen, and R. W. Mastwijk. 2002. *Manual for the Physical Anthropological Report*. Barge's Anthropologica 6. Leiden.

MacArthur, R., and E. O. Wilson. 1967. *The theory of island biogeography*. Princeton: Princeton University Press.

Macaulay, V., C. Hill, A. Achilli, C. Rengo, D. Clarke, W. Meehan, J. Blackburn, et al. 2005. Single, rapid coastal settlement of Asia revealed by analysis of complete mitochondrial genomes. *Science* 308 (5724): 1034–36.

Maclachlan, M. D., and W. F. Keegan. 1990. Archeology and the ethno-tyrannies. *American Anthropologist* 92: 1011–13.

Maíz, E. 1984. Reconocimiento arqueológico preliminar de la cuenca hidrográfica del Río Jauco. In *Proceedings of the Eleventh Congress of the International Association for Caribbean Archaeology*, ed. A. G. Pantel, I. V. Aenas, and M. S. Obediente, 312–23. St. Maarten, Netherlands Antilles.

Maiz López, E. J. 2002. El sitio arqueológico Hernández Colon: Actividades subsistenciales de los antiguos habitantes del Valle del Rio Cerrillos–Bucaná, Ponce, Puerto Rico. Master's thesis, Estudios Puertorriqueños, Centro de Estudios Avanzados de Puerto Rico y el Caribe, San Juan.

Malhi, R. S., J. A. Eshleman, J. A. Greenberg, et al. 2002. The structure of diversity within New World mitochondrial DNA haplogroups: Implications for the prehistory of North America. *American Journal of Human Genetics* 70: 905–19.

Malhi, R. S., H. M. Mortensen, J. A. Eshleman, et al. 2003. Native American mtDNA prehistory in the American Southwest. *American Journal of Physical Anthropology* 120: 108–24.

Malhi, R. S., B. A. Schultz, and D. G. Smith. 2001. Distribution of mitochondrial DNA lineages among Native American tribes of northeastern North America. *Human Biology* 73: 17–55.

Mandryk, C. A., H. Josenhans, D. W. Fedje, et al. 2001. Late Quaternary paleoenvironments of northwestern North America: Implications for inland versus coastal migration routes. *Quaternary Science Reviews* 20: 301–14.

Mann, C. C. 2005. *1491: New revelations of the Americas before Columbus*. New York: Random House.

Marichal García, L. 1995. Bibliografía arqueologica Cubana. In *Taíno: Arqueologia de Cuba*. CD-ROM. Colima: Centro de Antropología y CEDISAC.

Martin, P. S., and D. W. Steadman. 1999. Prehistoric extinctions on islands and continents. In *Extinctions in near time: Causes, contexts and consequences*, ed. R. D. MacPhee, 17–55. New York: Plenum Press.

Martin, R. 1956. *Lehrbuch der Anthropologie*, vol. 3, ed. K. Saller. 3rd ed. Stuttgart: Gustav Fischer.

Martínez-Cruzado, J. C. 2002. The use of mitochondrial DNA to discover pre-Columbian migrations to the Caribbean: Results for Puerto Rico and expectations for the Dominican Republic. *Kacike* 1–11.

Martínez-Cruzado, J. C., G. Toro-Labrador, V. Ho-Fung, M. Estévez-Montero, A. Lobaina-Manzanet, D. A. Padovani-Claudio, H. Sánchez-Cruz, P. Ortiz-Bermúdez, and A. Sánchez-Crespo. 2001. Mitochondrial DNA analysis reveals substantial Native American ancestry in Puerto Rico. *Human Biology* 73 (4): 491–511.

Martínez-Cruzado, J. C., G. Toro-Labrador, J. Viera-Vera, M. Y. Rivera-Vega, J. Startek, M. Latorre-Esteves, A. Román-Colón, et al. 2005. Reconstructing the population history of Puerto Rico by means of mtDNA phylogeographic analysis. *American Journal of Physical Anthropology* 128 (1): 131–55.

Martínez Fuentes, A., C. Lalueza Fox, T. P. Gilbert, A. Lazo Valdivia, F. Callafell, and J. Bertranpetit. 2003. El poblamiento antiguo del Caribe: Análisis del ADN mitocondrial en preagroalfareros de la región occidental de Cuba. *Catauro: Revista Cubana de Antropología* 5 (8): 62–74.

Martinón-Torres, M., R. Valcárcel Rojas, J. Cooper, and T. Rehren. 2007. Metals, microanalysis and meaning: A study of metal objects excavated from the indigenous cemetary of El Chorro de Maíta, Cuba. *Journal of Archaeological Science* 34: 194–204.

McAvoy, J. M. 2005. Williamson, Conover, and Cactus Hill: The significant Clovis sites of the Nottoway River drainage, southeastern Virginia. Paper presented at the Clovis in the Southeast Conference, Columbia, S.C. http://www.clovisinthesoutheast.net/jmcavoy.html.

McDonnell, J. A. 1991. *The dispossession of the American Indian, 1887–1934*. Bloomington: Indiana University Press.

Meltzer, D. J. 1997. Monte Verde and the Pleistocene peopling of the Americas. *Science* 276: 754–55.

Merriwether, D. A., F. Rothhammer, and R. E. Ferrell. 1995. Distribution of the four founding lineage haplotypes in Native Americans suggests a single wave of migration for the New World. *American Journal of Physical Anthropology* 98: 411–30.

Mesa, N. R., M. C. Mondragon, I. D. Soto, et al. 2000. Autosomal, mtDNA, and Y-chromosome diversity in Amerinds: Pre- and post-Columbian patterns of gene flow in South America. *American Journal of Human Genetics* 67: 1277–86.

Meyer, S., G. Weiss, and A. von Haeseler. 1999. Pattern of nucleotide substitution and rate heterogeneity in the hypervariable regions I and II of human mtDNA. *Genetics* 152 (3): 1103–10.

Mielke, J. E., and A. Long. 1969. Smithsonian Institution radiocarbon measurements. *Radiocarbon* 11 (1): 163–82.

Miyata, T., H. Hayashida, R. Kikuno, M. Hasegawa, M. Kobayashi, and K. Koike. 1982. Molecular clock of silent substitution: At least six-fold preponderance of silent changes in mitochondrial genes over those in nuclear genes. *Journal of Molecular Evolution* 19 (1): 28–35.

Monsalve, M. V., and E. Hagelberg. 1997. Mitochondrial DNA polymorphisms in Carib people of Belize. *Proceedings Royal Society of London, Biology* 264: 1217–24.

Moore, J. 2001. Evaluating five models of human colonization. *American Anthropologist* 103: 395–408.

Moore-Jansen, P. M., S. D. Ousley, and R. L. Jantz. 1994. *Data collection procedures for forensic skeletal material.* Report Investigation 48. Knoxville: University of Tennessee, Department of Anthropology.

Moreira de Lima, L. 1999. *La sociedad comunitaria de Cuba.* Havana: Felix Varela.

Mosimann, J. E., and F. C. James. 1979. New statistical methods for allometry with application to Florida red-winged blackbirds. *Evolution* 33: 444–59.

Murdock, G. P. 1949. *Social structure.* New York: Macmillan.

Narganes, Y. 1991. Secuencia cronológica de dos sitios arqueológicos de Puerto Rico (Sorcé, Vieques y Tecla, Guayanilla). In *Proceedings of the XIIIth Congress of the International Association for Caribbean Archaeology,* ed. E. N. Ayusi and J. B. Haviser, 628–46. Report of the Archaeological-Anthropological Institute of the Netherland Antilles 9. Curaçao, Netherland Antilles.

Narganes, Y. 2005. Nueva cronología de varios sitios de Puerto Rico y Vieques. Paper presented at the 21st International Congress for Caribbean Archaeology, Trinidad-Tobago.

Navarrete Pujol, R. 1990. *Caimanes III: Arqueología.* Havana: Editorial de Ciencias Sociales.

Neves, W. A., M. Hubbe, and G. Correal. 2007. Human skeletal remains from Sabana de Bogotá, Colombia: A case of Paleoamerican morphology late survival in South America? *American Journal of Physical Anthropology* 133 (4): 108–98.

Neves, W. A., and H. M. Pucciarelli. 1991. Morphological affinities of the first Americans: An exploratory analysis based on early South American human remains. *Journal of Human Evolution* 21: 261–73.

Newsom, L. 1993. Native West Indian plant use. Ph.D. diss., University of Florida, Gainesville.

Newsom, L. A., and E. Wing. 2004. *On land and sea: Native American uses of biological resources in the West Indies.* Tuscaloosa: University of Alabama Press.

Nicholson, D. V. 1976. An Antigua shell midden with Ceramic and Archaic components. In *Proceedings of the Sixth International Congress for the Study of Pre-Columbian Cultures of the Lesser Antilles,* ed. R. P. Bullen, 258–63. Gainesville: Florida State Museum and University of Florida.

Nuñez Jiménez, A. 1992. A 499 años de la Llegada de Colón. In *El V centenario visto desde Cuba,* ed. A. Hart Dávalos, A. Nuñez Jiménez, and S. Vilaseca Forné, 9–17. Holguín: Ediciones Holguín.

Oliver, J. R. 1999. The "La Hueca problem" in Puerto Rico and the Caribbean: Old problems, new perspectives, possible solutions. In *Archaeological investigations on St. Martín (Lesser Antilles),* ed. C. L. Hofman and M. L. P. Hoogland, 253–97. Archaeological Studies of Leiden University 4. Netherlands.

Oliver, J. R. 2006. Review of *People of the Caribbean: Cuba and Puerto Rico. Antiquity* 80 (308): 475–77.

Olsen, F. 1973. Did the Ciboney precede the Arawaks in Antigua? In *Proceedings of the*

*Fourth International Congress for the Study of Pre-Columbian Cultures of the Lesser Antilles*, 95–102. St. Lucia: St. Lucia Archaeological and Historical Society.

O'Rourke, D. H., M. G. Hayes, and S. W. Carlyle. 2000a. Ancient DNA studies in physical anthropology. *Annual Review of Anthropology* 29: 217–42.

O'Rourke, D. H., M. G. Hayes, and S. W. Carlyle. 2000b. Spatial and temporal stability of mtDNA haplogroup frequencies in native North America. *Human Biology* 72: 15–34.

Ortega, E., G. Atiles, and J. Ulloa. 2004. Investigaciones arqueológicas en el yacimiento La Iglesia, Provincia de Altagracia, República Dominicana. *Caribe Arqueológico* 8: 103–14.

Pagán Jiménez, J. R., M. Rodríguez, L. A. Chanlatte, and I. Narganes. 2005. La temprana introducción y uso de algunas plantas domésticas, silvestres y cultivos en Las Antillas precolombinas. *Diálogo Antropológico* 3 (10): 1–27.

Pauketat, T. 2001. Practice and history in archaeology: An emerging paradigm. *Anthropological Theory* 1: 733–98.

Pazdur, A., R. Awsiuk, A. Bluszcz, M. F. Pazdur, A. Walanus, and A. Zastawny. 1982. Gliwice radiocarbon dates VII. *Radiocarbon* 24 (2): 171–81.

Pendergast, D., E. Graham, J. Calvera, and J. Jardines. 2002. The houses in which they dwelt: The excavation and dating of Taino wooden structures at Los Buchillones, Cuba. *Journal of Wetland Archaeology* 2: 61–75.

Perego, U. A., A. Achilli, N. Angerhofer, M. Accetturo, M. Pala, A. Olivieri, B. H. Kashani, et al. 2009. Distinctive Paleo-Indian migration routes from Beringia marked by two rare mtDNA haplogroups. *Current Biology* 19: 1–8.

Perez, S. I., V. Bernal, and P. N. Gonzalez. 2007. Morphological differentiation of aboriginal human populations from Tierra del Fuego (Patagonia): Implications for South American peopling. *American Journal of Physical Anthropology* 133 (4): 1067–79.

Petersen, J., C. Hofman, and L. A. Curet. 2004. Time and culture: Chronology and taxonomy in the eastern Caribbean and the Guianas. In Delpuech and Hofman, *Late Ceramic Age societies in the eastern Caribbean*, 17–32.

Phillips, P., and G. R. Willey. 1953. Method and theory in American archaeology: An operational basis for culture-historical integration. *American Anthropologist* 55 (5): 615–33.

Pickerill, R. K., S. K. Donovan, and R. W. Portell. 2001. The Bioerosional ichnofossil *Petroxestes pera* Wilson and Palmer from the Middle Miocene of Carriacou. *Caribbean Journal of Science* 37: 130–31.

Pickerill, R. K., S. K. Donovan, and R. W. Portell. 2002. Bioerosional trace fossils from the Miocene of Carriacou, Lesser Antilles. *Caribbean Journal of Science* 38: 106–17.

Pino, M. 1995. *Actualizacion de fechados radiocarbónicos de sitios arqueológicos de Cuba hasta diciembre de 1993.* Havana: Editorial Academia.

Pino, M., and N. Castellanos. 1985. *Reporte de investigación 4.* Havana: Instituto de Ciencias Sociales.

Plog, S., and J. L. Hantman. 1990. Chronology construction and the study of prehistoric culture change. *Journal of Field Archaeology* 17 (4): 439–56.

Powell, J. F., and W. A. Neves. 1999. Craniofacial morphology of the first Americans:

Pattern and process in the peopling of the New World. *Yearbook of Physical Anthropology* 42: 153–88.

Pred, A. 1985. The social becomes spatial, the spatial becomes social: Enclosures, social change and the becoming of places in Skane. In *Social relations and spatial structures*, ed. D. Greggory and J. Urry, 337–65. New York: St. Martins Press.

Price, T. D., J. H. Burton, and R. A. Bentley. 2002. The characterisation of biologically available strontium isotope ratios for the study of prehistoric migration. *Archaeometry* 44: 117–35.

Price, T. D., C. M. Johnson, J. Ezzo, J. Ericson, and J. H. Burton. 1994. Residential mobility in the prehistoric Southwest United States: A preliminary study using strontium isotope analysis. *Journal of Archaeological Science* 21: 315–30.

Price, T. D., L. Manzanilla, and W. H. Middleton. 2000. Residential mobility at Teotihuacan: A preliminary study using strontium isotopes. *Journal of Archaeological Science* 27: 903–14.

Quintana-Murci, L., R. Veitia, M. Fellous, O. Semino, and E. S. Poloni. 2003. Genetic structure of Mediterranean populations revealed by Y-chromosome haplotype analysis. *American Journal of Physical Anthropology* 121 (2): 157–71.

Rainbird, P. 1999. Islands out of time: Toward a critique of island archaeology. *Journal of Mediterranean Archaeology* 12: 216–60.

Rainey, F. 1940. Scientific survey of Porto Rico and the Virgin Islands. *New York Academy of Sciences* 18, part I.

Rankin Santander, A. F. 1994. Estudio del sitio arqueológico de "Cabagán," Circuito Sur, Provincia de Sancti Spíritus. In *Estudios arqueologicos*, ed. J. Febles, L. Dominguez, F. Ortega, G. La Rosa, A. Martinez, and A. Rives, 129–39. Havana: Editorial Academia.

Reimer, P. 2005. Marine Reservoir Correction Database. http://radiocarbon.pa.qub.ac.uk/marine/ (last accessed April 14, 2005; updated February 17, 2005).

Reimer, P., F. G. McCormac, J. Moore, F. McCormick, and E. V. Murray. 2002. Marine radiocarbon reservoir corrections for the mid- to late Holocene in the eastern subpolar North Atlantic. *Holocene* 12 (2): 129–35.

Reimer, P. J., M. G. L. Baillie, E. Bard, A. Bayliss, J. W. Beck, C. J. H. Bertrand, P. G. Blackwell, et al. 2004. IntCal04 terrestrial radiocarbon age calibration, 0–26 cal kyr BP. *Radiocarbon* 46 (3): 1029–58.

Richard, G. 1994. Premier indice d'une occupation précéramique en Guadeloupe continentale. *Journal de la Société des Américanistes* 80: 241–42.

Rick, T. C., R. L. Vellanoweth, and M. J. Erlandson. 2005. Radiocarbon dating and the "old shell" problem: Direct dating of artifacts and cultural chronologies in coastal and other aquatic regions. *Journal of Archaeological Science* 32: 1641–48.

Rives Pantoja, A. V., J. Febles Duenas, M. E. Durán, A. Martinez, and L. Dominguez. 1991. *Los sitios arqueológicos de Cuba hasta 1990: Aplicaciones de la computación electrónica*. Havana: Centro de Antropología y Centro de Disenos de Sistema Automatizado, Academia de Ciencias de Cuba.

Robertson, R. 2005a. Grenada. In Lindsay et al., *Volcanic hazard atlas of the Lesser Antilles*, 50–61.

Robertson, R. 2005b. St. Vincent. In Lindsay et al., *Volcanic hazard atlas of the Lesser Antilles*, 241–61.

Robertson, R. 2005c. St. Kitts. In Lindsay et al., *Volcanic hazard atlas of the Lesser Antilles*, 205–17.

Robinson, L. S., E. Lundberg, and J. B. Walker, eds. 1985. Archaeological data recovery at El Bronce, Puerto Rico—final report, phase II. Archaeological Services, Inc., contract report submitted to the U.S. Army Corps of Engineers, Jacksonville. On file at the State Historic Preservation Office, San Juan, Puerto Rico.

Rodríguez López, M. 1984. *Cultural Resources Survey at Camp Santiago Salinas, Puerto Rico*. Prepared for the Puerto Rico National Guard in satisfaction of the requirements *of contract CX 5000-2-0886*. Copies available at the State Historic Preservation Office, San Juan, Puerto Rico.

Rodríguez López, M. A. 1990. Arqueología del Río Loiza. In *Proceedings of the 11th Congress of the International Association for Caribbean Archaeology*, ed. G. Pantel, 287–94. San Juan: Fundación Arqueológica e Historica de Puerto Rico.

Rodríguez López, M. A. 1991a. Arqueología de Punta Candelero, Puerto Rico. In *Proceedings of the XIIIth International Congress for Caribbean Archaeology*, ed. E. N. Ayubi and J. B. Haviser, 605–27. Reports of the Archaeological-Anthropological Institute of the Netherlands Antilles. Willemstad: Archaeological-Anthropological Institute of the Netherlands Antilles.

Rodríguez López, M. A. 1991b. Early trade networks in the Caribbean. In *Proceedings of the XIVth International Congress for Caribbean Archaeology*, ed. A. Cummins and P. King, 306–14. Bridgetown: Barbados Museum and Historical Society.

Rodríguez López, M. A. 2004. Excavaciones en el yacimiento Arcaico de Maruca, Ponce, Puerto Rico: Informe final. Copies available at the Consejo para la Protección del Patrimonio Arqueológico Terrestre de Puerto Rico, San Juan.

Rodríguez Ramos, R. 2001. Lithic reduction trajectories at La Hueca and Punta Candelero sites, Puerto Rico. Masters thesis, Texas A&M University.

Rodríguez Ramos, R. 2005. The crab-shell dichotomy revisited: The lithics speak out. In *Ancient Borinquen: Archaeology and ethnohistory of Native Puerto Rico*, ed. Peter Siegel, 1–54. Tuscaloosa: University of Alabama Press.

Rodríguez Ramos, R. 2006. Vertical and horizontal interactions in the precolonial Caribbean. Paper presented at the 71st Meetings of the Society for American Archaeology, San Juan, Puerto Rico.

Rodríguez Ramos, R. 2007. Puerto Rican precolonial history etched in stone. Ph.D. diss., University of Florida, Gainesville.

Rodríguez Ramos, R. 2008. From the Guanahatabey to the "Archaic" of Puerto Rico: The non-evident evidence. *Ethnohistory* 55 (3): 393–415.

Rodríguez Ramos, R., E. Babilonia, L. A. Curet, and J. Ulloa. 2008. The Pre-Arawak pottery horizon in the Antilles: A new approximation. *Latin American Antiquity* 19: 47–63.

Roe, P. G. 1981. Art and residence among the Shipibo Indians: A study in microacculturation. *American Anthropologist* 82: 42–71.

Roe, P. G. 1989. A grammatical analysis of Cedrosan Saladoid Vessel form categories and surface decoration: Aesthetic and technical styles in early Antillean ceramics. In *Early ceramic population lifeways and adaptive strategies in the Caribbean*, ed. P. E. Siegel, 267–382. BAR International Series 506. Oxford: British Archaeological Reports.

Roe, P. G. 1995. Eternal companions: Amerindian dogs from Terras Firma to the Antilles. In *Proceedings of the XVth International Congress for Caribbean Archaeology* [Puerto Rico 1993], ed. R. E. Alegría and M. Rodriguez, 155–72. San Juan: Centro de Estudios Avanzados de Puerto Rico y el Caribe.

Roosevelt, A. C., J. Douglas, and L. Brown. 2002. The migrations and adaptations of the first Americans: Clovis and pre-Clovis viewed from South America. In *The first Americans: The Pleistocene colonization of the New World*, ed. N. G. Jablonski, 159–235. San Francisco: California Academy of Sciences.

Rootsi, S., C. Magri, T. Kivisild, G. Benuzzi, H. Help, M. Bermisheva, I. Kutuev, et al. 2004. Phylogeography of Y-chromosome haplogroup I reveals distinct domains of prehistoric gene flow in Europe. *American Journal of Human Genetics* 75 (1): 128–37.

Ross, A. H. 2004. Cranial evidence of pre-contact multiple population expansions in the Caribbean. *Caribbean Journal of Science* 40 (3): 291–98.

Ross, A. H., D. H. Ubelaker, and A. B. Falsetti. 2002a. Craniometric variation in the Americas. *Human Biology* 74: 807–18.

Ross, A. H., D. H. Ubelaker, and A. B. Falsetti. 2002b. Ethnohistorical relationships on the Iberian Peninsula. *Anthropologie* 40 (1): 51–57.

Ross, A. H., D. H. Ubelaker, and S. Guillén. 2008. Craniometric patterning within ancient Peru. *Latin American Antiquity* 19 (2): 158–66.

Roth, W. E. 1924. *An introductory study of the arts, crafts, and customs of the Guiana Indians*. Bureau of American Ethnology to the Secretary of the Smithsonian Institution, Thirty-eighth Annual Report. Washington, D.C.: Government Printing Office.

Rothhammer, F., J. A. Cocilovo, S. Quevedo, and E. Llop. 1982. Microevolution in prehistoric Andean populations: Chronologic craniometric variation. *American Journal of Physical Anthropology* 58 (4): 391–96.

Rothhammer, F., J. A. Cocilovo, S. Quevedo, and E. Llop. 1985. The settlement of South-America. *Archivos de Biologia y Medicina Experimentales* 18: R73.

Rothhammer, F., S. Quevedo, J. A. Cocilovo, and E. Llop. 1984. Microevolution in prehistoric Andean populations: Chronologic nonmetrical cranial variation in northern Chile. *American Journal of Physical Anthropology* 65 (2): 157–62.

Rouse, I. R. 1952. Porto Rican prehistory: Introduction; Excavations in the west and north. In *Scientific Survey of Porto Rico and the Virgin Islands*, vol. 18, part 3. New York: New York Academy of Sciences.

Rouse, I. R. 1963. *Final technical report (NSF-G24049): Dating of Caribbean cultures*. New Haven, Conn.: Department of Anthropology, Yale University.

Rouse, I. 1976. The Saladoid sequence on Antigua and its aftermath. In *Proceedings of the Sixth International Congress for the Study of Pre-Columbian Cultures of the Lesser Antilles*, ed. R. P. Bullen, 35–41. Gainesville: Florida State Museum.

Rouse, I. R. 1986. *Migrations in prehistory: Inferring population movement from cultural remains*. New Haven, Conn.: Yale University Press.

Rouse, I. 1989. Peoples and cultures of the Saladoid frontier in the Greater Antilles. In

*Early Ceramic population lifeways and adaptive strategies in the Caribbean,* ed. P. E. Siegel, 383–403. BAR International Series 506. Oxford: British Archaeological Reports.

Rouse, I. 1992a. *The Tainos: Rise and decline of the people who greeted Columbus.* New Haven, Conn.: Yale University Press.

Rouse, I. 1992b. West Indian chronological and cultural systems and their use in the Leeward Islands. Paper presented at 57th Meetings of the Society for American Archaeology, Pittsburgh.

Rouse, I., and R. E. Alegría. 1979. Radiocarbon dates from the West Indies. *Review Interamericana* 8 (3): 495–99.

Rouse, I., and R. E. Alegría. 1990. *Excavations at María de la Cruz Cave and Hacienda Grande Village Site, Loiza, Puerto Rico.* Yale University Publications in Anthropology. New Haven, Conn.: Yale University Press.

Rouse, I. and L. Allaire. 1978. Caribbean. In *Chronologies of the New World,* ed. R. E. Taylor and C. W. Meighan, 432–81. New York: Academic Press.

Rubel, P. G., and A. Rosman. n.d. Solidarity of the sibling group: Cognatic societies and their mortuary remains. Manuscript in possession of the authors, cited with permission.

Ruhlen, M. 1994. Linguistic evidence for the peopling of the Americas. In *Method and theory for investigating the peopling of the Americas,* ed. R. Bonnichsen and D. G. Steele, 177–88. Peopling of the Americas Publications. Corvallis: Center for the Study of the First Americans, Department of Anthropology, Oregon State University.

Saiki, R. K., S. Scharf, F. Faloona, K. B. Mullis, G. T. Horn, H. A. Erlich, and N. Arnheim. 1985. Enzymatic amplification of beta-globin genomic sequences and restriction site analysis for diagnosis of sickle cell anemia. *Science* 230 (4732): 1350–54.

Saillard, J., P. Forster, N. Lynnerup, H. J. Bandelt, and S. Norby. 2000. mtDNA variation among Greenland Eskimos: The edge of the Beringian expansion. *American Journal of Human Genetics* 67 (3): 718–26.

Salas, A., M. Richards, M. V. Lareu, et al. 2005. Shipwrecks and founder effects: Divergent demographic histories reflected in Caribbean mtDNA. *American Journal of Physical Anthropology* 128: 855–60.

Salzano, F. M. 2002. Molecular variability in Amerindians: Widespread but uneven information. *Anais da Academia Brasileira de Ciencas* 74 (2): 223–63.

Salzano, F. M., and S. M. Callegari-Jacques. 1988. *South American Indians: A case study in evolution.* New York: Oxford University Press.

Santos, F. R., N. O. Bianchi, and S. D. J. Pena. 1996. Worldwide distribution of human Y-chromosome haplotypes. *Genome Research* 6 (7): 601–11.

Santos, M. R., R. H. Ward, and R. Barrantes. 1994. mtDNA variation in the Chibcha Amerindian Huetar from Costa Rica. *Human Biology* 66: 963–77.

Santos, S. E. B., A. K. C. Ribeiro-Dos-Santos, D. Meyer, et al. 1996. Multiple founder haplotypes of mitochondrial DNA in Amerindians revealed by RFLP and sequencing. *Annals of Human Genetics* 60: 305–19.

Sardi Marina L., F. Ramirez Rozzi, R. Gonzalez-Jose, and H. M. Pucciarelli. 2005. South

Amerindian craniofacial morphology: Diversity and implications for Amerindian evolution. *American Journal of Physical Anthropology* 128: 747–56.

SAS. System for Windows, version 9.1.3 [computer program]. Cary, N.C.

Schurr, T. G. 2004a. The peopling of the New World: Perspectives from molecular anthropology. *Annual Review of Anthropology* 33 (3): 551–83.

Schurr, T. G., S. W. Ballinger, Y. Y. Gan, J. A. Hodge, D. A. Merriwether, D. N. Lawrence, W. C. Knowler, K. M. Weiss, and D. C. Wallace. 1990. Amerindian mitochondrial DNAs have rare Asian mutations at high frequencies suggesting they derived from four primary maternal lineages. *American Journal of Human Genetics* 46 (3): 613–23.

Schurr, T. G., and S. T. Sherry. 2004. Mitochondrial DNA and Y chromosome diversity and the peopling of the Americas: Evolutionary and demographic evidence. *American Journal of Human Biology* 16 (4): 420–39.

Schurr, T. G., R. I. Sukernik, Y. B. Starikovsdaya, and D. C. Wallace. 1999. Mitochondrial DNA variation in the Okhotsk Sea–Bering region during the Neolithic. *American Journal of Physical Anthropology* 110: 271–84.

Schurr, T. G., and D. C. Wallace. 1999. mtDNA variation in Native Americans and Siberians and its implications for the peopling of the New World. In *Who were the first Americans?* ed. R. Bonnichsen, 41–77. Corvallis: Center for the Study of the First Americans, Department of Anthropology, Oregon State University.

Schweissing, M. M., and G. Grupe. 2003. Stable strontium isotopes in human teeth and bone: A key to migration events of the late Roman period in Bavaria. *Journal of Archaeological Science* 30: 1373–83.

Seielstad, M. T., N. Yuldasheva, N. Singh, P. Underhill, P. Oefner, P. Shen, and R. S. Wells. 2003. A novel Y-chromosome variant puts an upper limit on the timing of first entry into the Americas. *American Journal of Human Genetics* 73 (3): 700–705.

Serrand, N. 1999. Long distance transportation of freshwater bivalves (Unionidae) in the Lesser Antilles during the 1st millennium AD: Example from the Hope Estate Saladoid site (St. Martin). In *Proceedings of the XVIIIth International Congress for Caribbean Archaeology I* [Grenada 1999], 136–52. Guadeloupe: International Association for Caribbean Archaeology.

Siegel, P., ed. 1989. *Early Ceramic population lifeways and adaptive strategies in the Caribbean*. BAR International Series 506. Oxford: British Archaeological Reports.

Siegel, P. E. 1991. Migration research in Saladoid archaeology: A review. *Florida Anthropologist* 44: 79–91.

Siegel, P. E. 1996. Ideology and culture change in prehistoric Puerto Rico: A view from the community. *Journal of Field Archaeology* 23: 313–33.

Siegel, P., ed. 2005. *Ancient Borinquen: Archaeology and ethnohistory of Native Puerto Rico*. Tuscaloosa: University of Alabama Press.

Sigurdsson, H., and S. Carey. 1991. *Caribbean volcanoes: A field guide*. Toronto: Geological Association of Canada.

Silva, W. A., S. L. Bonatto, A. J. Holanda, et al. 2002. Mitochondrial genome diversity of Native Americans supports a single early entry of founder populations into America. *American Journal of Human Genetics* 71: 187–92.

Simpson, K. 2005. Nevis. In Lindsay et al., *Volcanic hazard atlas of the Lesser Antilles*, 169–90.

Slice, D. E. 1998. Morpheus et al. [computer program]. http://life.bio.sunysb.edu/morph/main.html.

Smith, A., and M. Roobol. 2005a. Saba. In Lindsay et al., *Volcanic hazard atlas of the Lesser Antilles*, 181–91.

Smith, A., and M. Roobol. 2005b. St. Eustatius. In Lindsay et al., *Volcanic hazard atlas of the Lesser Antilles*, 192–203.

Sneath, P. H. A., and R. R. Sokal. 1973. *Numerical taxonomy.* London: Freeman.

Soares, P., L. Ermini, N. Thomson, M. Mormina, T. Rito, A. Röhl, A. Salas, S. Oppenheimer, V. Macaulay, and M. B. Richards. 2009. Correcting for purifying selection: An improved human mitochondrial molecular clock. *American Journal of Human Genetics* 84 (6): 740–59.

Sokal, R. R., G. M. Jacquez, N. L. Oden, D. DiGiovanni, A. B. Falsetti, E. McGee, and B. A. Thomson. 1993. Genetic relationships of European populations reflect their ethnohistorical affinities. *American Journal of Physical Anthropology* 91: 55–70.

Spriggs, Matthew. 1989. The dating of the Island S.E. Asian Neolithic: An attempt at chronometric hygiene and linguistic correlation. *Antiquity* 63: 587–613.

Steadman, D. W., P. S. Martin, R. D. E. MacPhee, A. J. T. Jull, H. Gregory McDonald, C. A. Woods, M. Iturralde-Vinent, and G. W. L. Hodgins. 2005. Asynchronous extinction of late quaternary sloths on continents and islands. *Proceedings of the National Academy of Sciences* 102 (33): 11763–68.

Steele, D. G., and J. F. Powell. 1993. Paleobiology of the first Americans. *Evolutionary Anthropology* 2: 138–46.

Steele, D. G., and J. F. Powell. 1994. Paleobiological evidence for the peopling of the Americas: A morphometric view. In *Method and theory for investigating the peopling of the Americas*, ed. R. Bonnichsen and D. G. Steele, 141–64. Peopling of the Americas Publications. Corvallis: Center for the Study of the First Americans, Department of Anthropology, Oregon State University.

Stokstad, E. 2000. "Pre-Clovis" site fights for recognition. *Science* 288 (5464): 247.

Stone, A., and M. Stoneking. 1998. mtDNA analysis of a prehistoric Oneota population: Implications for the peopling of the New World. *American Journal of Human Genetics* 62: 1153–70.

Stone, W., and A. Hays. 1991. *A cruising guide to the Caribbean.* New York: Putman.

Stuckenrath, R., and J. E. Mielke. 1973. Smithsonian Institution radiocarbon measurements VIII. *Radiocarbon* 15 (2): 388–424.

Stuiver, M., and T. F. Braziunas. 1993. Modeling atmospheric $^{14}$C influences and $^{14}$C ages of marine samples to 10,000 BC. *Radiocarbon* 35 (1): 137–89.

Stuiver, M., and H. A. Polach. 1977. Reporting of $^{14}$C data—discussion. *Radiocarbon* 19 (3): 355–63.

Stuiver, M., and P. Reimer. 1993. Extended $^{14}$C database and Revised CALIB 3.0 $^{14}$C age calibration program. *Radiocarbon* 35: 215–30.

Stuiver, M., B. J. Reimer, and J. W. Reimer. 2005. CALIB 5.0 (online program and documentation). http://intcal.qub.ac.uk/calib/.

Stuiver, M., P. J. Reimer, E. Bard, J. W. Beck, G. S. Burr, K. A. Hughen, B. Kromer, G. McCormac, J. van der Plicht, and M. Spurk. 1998. INTCAL98 radiocarbon age calibration, 24,000–0 BP. *Radiocarbon* 40 (3): 1041–83.

Sutty, L. 1990. A listing of Amerindian settlements on the island of Carriacou in the southern Grenadines and a report on the most important of these, Grand Bay. In *Proceedings of the 11th Congress of the International Association for Caribbean Archaeology,* ed. P. Tekakis, A. Gus, I. Vargas Arenas, and M. S. Obediente, 242–59. San Juan: Fundación Arqueológica, Antropológica e Historica de Puerto Rico.

Tabio, E. 1974. La comunidad primitiva. *Revolución y Cultura Septiembre,* 1: 7–14.

Tabio, E. 1984. Nueva periodización para el estudio de las comunidades aborígenes de Cuba. *Islas* 78: 37–52.

Tabio, E. E. 1995. *Introducción a la arqueología de las Antillas.* Havana: Editorial de Ciencias Sociales.

Tabio, E. E., and J. M. Guarch. 1966. *Excavaciones en Arroyo del Palo, Mayari, Cuba.* Havana: Academia de Ciencias de la República de Cuba.

Tabio, E., and E. Rey. 1979. *Prehistoria de Cuba.* 2nd ed. Havana: Editorial de Ciencias Sociales.

Tajima, A., K. Hamaguchi, H. Terao, et al. 2004. Genetic background of people in the Dominican Republic with or without obese type 2 diabetes revealed by mitochondrial DNA polymorphism. *Journal of Human Genetics* 49: 495–99.

Tajima, F. 1989. Statistical method for testing the neutral mutation hypothesis by DNA polymorphism. *Genetics* 123 (3): 585–95.

Takamiya, H. 2006. An unusual case? Hunter-gatherer adaptations to an island environment: A case study from Okinawa, Japan. *Journal of Island and Coastal Archaeology* 1 (1): 49–66.

Tamm, E., T. Kivisild, M. Reidla, M. Metspalu, D. G. Smith, C. J. Mulligan, C. M. Bravi, et al. 2007. Beringian standstill and spread of Native American founders. *PLoS ONE* 2 (9): e829.

Taylor, R. E. 1987. *Radiocarbon dating: An archaeological perspective.* London: Academic Press.

Thomas, H. L., and R. W. Ehrich. 1969. Some problems with chronology. *World Archaeology* 1 (2): 143–56.

Thornton, R., and J. Marsh-Thornton. 1981. Estimating prehistoric American Indian population size for United States area: Implications of the nineteenth century decline and nadir. *American Journal of Physical Anthropology* 55: 47–53.

Toro-Labrador, G., O. Wever, and C. Martínez-Cruzado. 2003. Mitochondrial DNA analysis in Aruba: Strong maternal ancestry of closely related Amerindians and implications for the peopling of northwestern Venezuela. *Caribbean Journal of Science* 39: 11–22.

Torres, J. M. 2000. Settlement patterns and political geography of the Saladoid and Ostionoid peoples of south-central Puerto Rico. Master's thesis, University of Colorado, Denver.

Torres, J. M. 2005. Deconstructing the polity: Communities and social landscapes of the Ceramic Age peoples of south central Puerto Rico. In *Ancient Borinquen: Archae-*

*ology and ethnohistory of Native Puerto Rico*, ed. Peter Siegel, 202–29. Tuscaloosa: University of Alabama Press.

Torres Etayo, D. 2004. La arqueología cubana en la encrucijada: La teoría o la empiria. *Caribe Arqueológico* 8: 2–7.

Torroni, A., H. J. Bandelt, L. D'Urbano, P. Lahermo, P. Moral, D. Sellitto, C. Rengo, et al. 1998. mtDNA analysis reveals a major late Paleolithic population expansion from southwestern to northeastern Europe. *American Journal of Human Genetics* 62 (5): 1137–52.

Torroni, A., M. D. Brown, M. T. Lott, et al. 1995. African, Native American and European mitochondrial DNAs in Cubans from Pinar del Rio province and implications for the recent epidemic neuropathy in Cuba. *Human Mutation* 5: 310–17.

Torroni, A., J. V. Neel, R. Barrantes, T. G. Schurr, and D. C. Wallace. 1994. Mitochondrial DNA "clock" for the Amerinds with its implications for timing their entry into North America. *Proceedings of the National Academy of Sciences USA* 91 (3): 1158–62.

Torroni, A., T. G. Schurr, M. F. Cabell, M. D. Brown, J. V. Neel, M. Larsen, D. G. Smith, C. M. Vullo, and D. C. Wallace. 1993. Asian affinities and continental radiation of the four founding Native American mtDNAs. *American Journal of Human Genetics* 53 (3): 563–90.

Torroni, A., T. G. Schurr, C. C. Yang, et al. 1992. Native American mitochondrial DNA analysis indicates that the Amerind and the NaDene populations were founded by two independent migrations. *Genetics* 130: 153–62.

Trincado Fontán, M. N. 2000. El aborigen, la historiografía y la nacionalidad cubanas. *Caribe Arqueologico* 4: 100–105.

Trincado Fontán, M. N., and J. Ulloa Hung. 1996. Las comunidades meillacoides del litoral sudoriental de Cuba. *Caribe Arqueológico* 1: 74–82.

Troche-Matos, M. I. 2001. Origen geográfico de los Dos Subhaplogrupos C identificados en la población actual Puertorriqueña mediante análisis de restricción en el DNA mitocondrial. M.S. thesis, University of Puerto Rico, Mayagüez.

Tronolone, C. A., M. A. Cinquino, and C. E. Vandrei. 1984. Cultural resource reconnaissance survey for the Vieques Naval Reservation. Copies available at the Puerto Rico State Historic Preservation Office, San Juan.

Turner, C. G. 1985. The dental search for Native American origins. In *Out of Asia: Peopling the Americas and the Pacific*, ed. R. Kirk and E. Szathmary, 31–78. Canberra: Journal of Pacific History.

Turner, C. G. 1994. Relating Eurasian and Native American populations through dental morphology. In *method and theory for investigating the peopling of the Americas*, ed. R. Bonnichsen and D. G. Steele, 131–40. Peopling of the Americas Publications. Corvallis: Center for the Study of the First Americans, Department of Anthropology, Oregon State University.

Ubelaker, D. H. 1981. The Ayalan cemetery: A late integration burial site on the south coast of Ecuador. *Smithsonian Contributions to Anthropology* 29. Washington, D.C.

Ubelaker, D. H. 1987. Dental alteration in prehistoric Ecuador: A new example from Jama-coaque. *Journal of the Washington Academy of Sciences* 77: 76–80.

Ubelaker, D. H. 1988. North American Indian population size, A.D. 1500 to 1985. *American Journal of Physical Anthropology* 77: 289–94.

Ubelaker, D. H., and R. L. Jantz. 1986. Biological history of the aboriginal population of North America. In *Rassengeschichte der Menschheit: 11. Lieferung: Amerika I: Nordamerika, Mexico,* ed. Ilse Schwidetzky, 7–79. Munich: R. Oldenbourg.

Ulloa Hung, J., and R. Valcárcel Rojas. 2002. *Ceramica temprana en el centro del oriente de Cuba.* Santo Domingo: View Graph Impresos.

Underhill, P. A., L. Jin, A. A. Lin, S. Q. Mehdi, T. Jenkins, D. Vollrath, R. W. Davis, L. L. Cavalli-Sforza, and P. J. Oefner. 1997. Detection of numerous Y chromosome biallelic polymorphisms by denaturing high-performance liquid chromatography. *Genome Research* 7 (10): 996–1005.

Underhill, P. A., L. Jin, R. Zemans, P. J. Oefner, and L. L. Cavalli-Sforza. 1996. A pre-Columbian Y chromosome–specific transition and its implications for human evolutionary history. *Proceedings of the National Academy of Sciences* 93 (1): 196–200.

Underhill, P. A., G. Passarino, A. A. Lin, P. Shen, M. M. Lahr, R. A.Foley, P. J. Oefner, and L. L. Cavalli-Sforza. 2001. The phylogeography of Y chromosome binary haplotypes and the origins of modern human populations. *Annals of Human Genetics* 65 (1): 43–62.

Underhill, P. A., P. Shen, A. A. Lin, L. Jin, G. Passarino, W. H. Yang, E. Kauffman, et al. 2000. Y chromosome sequence variation and the history of human populations. *Nature Genetics* 26 (3): 358–61.

United States Navy. 1995. *Marine Climatic Atlas of the World.* Asheville, N.C.: National Climatic Data Center.

Valcárcel Rojas, R. 2002. *Banes Precolombino: La ocupacion agricultora.* 1st ed. Holguín: Ediciones Holguín.

Valcárcel Rojas, R., J. Cooper, J. Calvera Rosés, O. Brito, and M. Labrada. 2006. Postes en el mar: Excavación de una estructura constructiva aborigen en Los Buchillones. *Caribe Arqueologico* 9: 76–88.

Varela, H. H., and J. A. Cocilovo. 1999. Evaluation of the environmental component of the phenotypic variance in prehistoric populations. *Homo* 50: 46–53.

Varela, H. H., and J. A. Cocilovo. 2000. Structure of the prehistoric population of San Pedro de Atacama. *Current Anthropology* 41: 125–32.

Varela, H. H., and J. A. Cocilovo. 2002. Genetic drift and gene flow in a prehistoric population of the Azapa Valley and Coast, Chile. *American Journal of Physical Anthropology* 118: 259–67.

Veloz Maggiollo, M., and B. Vega. 1982. The Antillean Preceramic: A new approximation. *Journal of New World Archaeology* 5: 33–44.

Versteeg, A. H., and K. Schinkel. 1992. *The archaeology of St. Eustatius: The Golden Rock site.* Amsterdam: Foundation for Scientific Research in the Caribbean Region.

Vescelius, G. S., and L. S. Robinson. 1979. Exotic items in archaeological collections from St. Croix: Prehistoric imports and their implications. Paper presented at the VIIIth International Congress for the Study of the Pre-Columbian Cultures of the Lesser Antilles, St. Kitts.

Vigilant, L., R. Pennington, H. Harpending, T. D. Kocher, and A. C. Wilson. 1989. Mitochondrial DNA sequences in single hairs from a southern African population. *Proceedings of the National Academy of Sciences USA* 86 (23): 9350–54.

Vinogradov, A. P., A. L. Devirts, E. I. Dobkina, and N. G. Markova. 1968. Radiocarbon dating in the Vernadsky Institute, Institute of Geochemistry and Analytical Chemistry, Academy of Sciences, USSR, Moscow. *Radiocarbon* 10 (2): 454–64.

Volodko, N. V., E. B. Starikovskaya, I. O. Mazunin, N. P. Eltsov, P. V. Naidenko, D. C. Wallace, and R. I. Sukernik. 2008. Mitochondrial genome diversity in Arctic Siberians, with particular reference to the evolutionary history of Beringia and Pleistocenic peopling of the Americas. *American Journal of Human Genetics* 82 (5): 1084–1100.

Wakeley, J. 1993. Substitution rate variation among sites in hypervariable region I of human mitochondrial DNA. *Journal of Molecular Evolution* 37 (6): 613–23.

Wallace, D. C., M. D. Brown, and M. T. Lott. 1999. Mitochondrial DNA variation in human evolution and disease. *Gene* 238: 211–30.

Ward, R. H., F. M. Salzano, S. L. Bonatto, M. H. Hutz, C. E. A. Coimbra, and R. V. Santos. 1996. Mitochondrial DNA polymorphism in three Brazilian Indian tribes. *American Journal of Human Biology* 8 (3): 317–23.

Waters, M. R., S. L. Forman, T. W. Stafford Jr., et al. 2005. Geoarchaeological investigations at the Topper and Big Pine sites. Paper presented at the Clovis in the Southeast Conference, Columbia, S.C. http://www.clovisinthesoutheast.net/waters.html.

Watson, E., P. Forster, M. Richards, and H. J. Bandelt. 1997. Mitochondrial footprints of human expansions in Africa. *American Journal of Human Genetics* 61 (3): 691–704.

Watters, D. R. 1997. Maritime trade in the prehistoric eastern Caribbean. In *The indigenous people of the Caribbean*, ed. S. Wilson, 88–99. Gainesville: University Press of Florida.

Watters, D. R., J. Donahue, and R. Stuckenrath. 1992. Paleoshorelines and the prehistory of Barbuda, West Indies. In *Paleoshorelines and prehistory*, ed. L. Johnson and M. Stright, 15–52. Boca Raton, Fla.: CRC Press.

Watters, D. R., and I. B. Rouse. 1989. Environmental diversity and maritime adaptations in the Caribbean area. In *Early Ceramic population lifeways and adaptive strategies in the Caribbean*, ed. P. E. Siegel, 129–44. BAR International Series 506. Oxford: British Archaeological Reports.

Watters, D. R., and R. Scaglion. 1994. Beads and pendants from Trants, Montserrat: Implications for the prehistoric lapidary industry of the Caribbean. *Annals of Carnegie Museum* 63: 215–37.

Weaver, G. G., P. H. Garrow, and J. R. Oliver. 1991. Phase II archaeological data recovery at PO-38, EL-Parking site, Barrio Maraguez, Ponce, Puerto Rico. Report submitted to the U.S. Army Engineer District. Copies available at the State Historic Preservation Office, San Juan.

Wilson, S. M. 1989. The prehistoric settlement pattern of Nevis, West Indies. *Journal of Field Archaeology* 16: 427–50.

Wilson, S. M., ed. 1997. *The indigenous people of the Caribbean*. Gainesville: University Press of Florida.

Wilson, S. M. 2007. *The archaeology of the Caribbean*. Cambridge: Cambridge University Press.

Wilson, S. M., H. B. Iceland, and T. R. Hester. 1998. Preceramic connections between Yucatan and the Caribbean. *Latin American Antiquity* 9 (4): 342–52.

Workshop of European Anthropologists (WEA). 1980. Recommendations for age and sex diagnoses of skeletons. *Journal of Human Evolution* 9: 517–49.

Wright, L. E. 2005. Identifying immigrants to Tikal, Guatemala: Defining local variability in strontium isotope ratios of human tooth enamel. *Journal of Archaeological Science* 32: 555–66.

Y Chromosome Consortium. 2002. A nomenclature system for the tree of human Y-chromosomal binary haplogroups. *Genome Research* 12 (2): 339–48.

Zegura, S. L., T. M. Karafet, L. A. Zhivotovsky, et al. 2004. High-resolution SNPs and microsatellite haplotypes point to a single, recent entry of Native American Y chromosomes into the Americas. *Molecular Biology and Evolution* 21: 164–75.

Zerjal, T., R. S. Wells, N. Yuldasheva, R. Ruzibakiev, and C. Tyler-Smith. 2002. A genetic landscape reshaped by recent events: Y-chromosomal insights into Central Asia. *American Journal of Human Genetics* 71 (3): 466–82.

Zhivotovsky, L. A. 2001. Estimating divergence time with the use of microsatellite genetic distances: Impacts of population growth and gene flow. *Molecular Biology and Evolution* 18 (5): 700–709.

Zhivotovsky, L. A., P. A. Underhill, C. Cinnioğlu, M. Kayser, B. Morar, T. Kivisild, R. Scozzari, et al. 2004. The effective mutation rate at Y chromosome short tandem repeats, with application to human population-divergence time. *American Journal of Human Genetics* 74 (1): 50–61.

# Contributors

Richard T. Callaghan is associate professor in the Department of Archaeology at the University of Calgary. He specializes in the archaeology of lowland South America and the Caribbean and the development of water transport and navigation. His recent fieldwork has been focused on prehistoric settlement patterns on St. Vincent in the eastern Caribbean. He has developed a computer simulation program to investigate ancient and historical voyaging around the world. Callaghan's research using computer simulations of ancient voyaging has included the Caribbean, the Pacific coast of Central and northern South America, Edo period contacts between Japan and the Northwest Coast of North America, early prehistoric contacts between the Sea of Okhotsk region and the Northwest Coast of North America, and Micronesia.

Jago Cooper is Leverhulme Early Career Fellow in the School of Archaeology and Ancient History at the University of Leicester. He is codirector of the Los Buchillones Research Project, a collaborative venture between archaeologists from the Cuban Ministry of Science Technology and Environment and the University College London Institute of Archaeology. His research interests include island archaeology, landscape survey, GIS applications, zooarchaeology, and Latin American archaeology. Cooper developed his interest in Caribbean archaeology working on projects in Mexico, Belize, Colombia, Barbados, and Puerto Rico before concentrating his doctoral research on Cuba. His study focuses on island interaction in the prehistoric Caribbean, building on an archaeological case study from the Sabana-Camaguey archipelago in north-central Cuba.

Scott M. Fitzpatrick is professor of archaeology at the University of Oregon who specializes in the archaeology of islands and coastal regions. His research interests include colonization strategies, interaction and exchange systems, maritime adaptations, site taphonomy, and historical ecology, particularly in the Pacific and Caribbean. He has authored over fifty scholarly papers, including recent contributions to *Coral Reefs*, *Geoarchaeology*, *Journal of Archaeological Science*, *Journal of Field Archaeology*, and *Latin American Antiquity* and edited *Voyages of Discovery: The Archaeology of Islands*. Fitzpatrick is also founder and coeditor of the *Journal of Island and Coastal Archaeology*.

Christina M. Giovas is a zooarchaeologist with research interests in the prehistory of islands, particularly those of the Caribbean and Polynesia. Her work focuses on understanding human-environmental relationships over time as informed by behavioral ecology and historical ecology frameworks. Her current doctoral research is focused on investigating the comparative foraging patterns of inhabitants at multiple prehistoric sites on the island of Carriacou, Grenada. Giovas is the 2006 recipient of the University of Washington Department of Anthropology's Peggy Yeager Award for scholarly excellence in archaeology.

Corinne L. Hofman is professor of Caribbean archaeology at Leiden University in the Netherlands. She has been conducting archaeological research on several islands in the Lesser Antilles and the Dominican Republic. Over the past few years, she was been awarded several grants from the Netherlands Organisation for Scientific Research (NWO) for innovational research in the Caribbean. Her current research themes center on ceramic analysis, mobility and exchange, and sociopolitical organization in the precolonial Caribbean. She is currently the chair of education on the faculty board.

Menno L. P. Hoogland is associate professor of archaeology at Leiden University. His research focuses on burial practices during the Late Ceramic Age and mobility within and between residential groups. The latter aspect relies on the analysis of strontium isotope ratios in human and faunal material. In 2004, he was awarded a grant by the Netherlands Organisation for Scientific Research (NWO) for the program "Houses for the Living and the Dead." The project focuses on the organization of Taíno households in the Dominican Republic (AD 1000–1500) and is executed in cooperation with the Museo del Hombre Dominicana. He is currently the director of the Undergraduate School.

Michiel Kappers is founder of In-Terris Site Technics, an archaeological contract firm specializing in the computerized acquisition and interpretation of data recovered in survey and excavation. His research interests include GIS, global positioning system survey techniques, and database software development. He has worked extensively on large-scale archaeological projects in the Netherlands, where he is involved with the development of a standardized national archaeological database system, and on several islands in the Caribbean, including Guadeloupe, Jamaica, Trinidad, and Carriacou.

William F. Keegan is curator of Caribbean archaeology at the Florida Museum of Natural History and professor of anthropology and Latin American studies at the University of Florida, Gainesville. He has conducted research in the West Indies since 1978 and has directed projects in the Bahamas, Turks and Caicos, Grand Cayman, Haiti, Jamaica, Cuba, Dominican Republic, Puerto Rico, Grenada, St. Lucia, and Trinidad. The results of his research are presented in five books (two edited), more than forty peer-reviewed scientific papers, and numerous popular publications. One of his recent projects was the investigation of the Taíno ceremonial center on Middle Caicos, Turks and Caicos Islands. Research at this site is the topic of his recently published book, *Taino Indian Myth and Practice: The Arrival of the Stranger King*.

Juan C. Martínez-Cruzado is professor of biology at the University of Puerto Rico, Mayagüez. He is a geneticist specializing in population genetics and molecular evolution. His research interests include mitochondrial DNA analysis of Caribbean populations and its revelations on ancient Amerindian migratory routes and demography. His recent publications include papers in the *American Journal of Physical Anthropology, Caribbean Journal of Science, Human Biology, Developmental Genetics, Journal of Molecular Evolution*, and *Theoretical and Applied Genetics*.

José R. Oliver is currently lecturer in Latin American archaeology at the Institute of Archaeology, University College London, where he teaches archaeology courses on the Andes, Amazonia, and the Caribbean. Since 1995, he has been codirector of the Utuado-Caguana Archaeological

Project in collaboration with the Instituto de Cultura Puertorriqueña and the Oficina Estatal de Conservación Histórica (PR-SHPO). His research interests include the origins of food production in Amazonia and the circum-Caribbean region and the nature and development of Antillean *cacicazgos*. He is author of *El centro ceremonial de Caguana* and *Caciques and Cemi Idols: The Web Spun by Taino Rulers between Hispaniola and Puerto Rico* and coeditor with Colin McEwan and Anna Casas of *El Caribe precolombino: Fray Ramón Pané y el universo taíno*, which accompanied a major Taíno exhibition in Barcelona, Santiago de Compostela, and Madrid in 2008–2009.

Raphaël G. A. M. Panhuysen is a research fellow in the Amsterdam Archaeological Centre at the University of Amsterdam. He is a project leader in the Dorestad *vicus famosus* project funded by the Netherlands Organisation for Scientific Research (NWO), aimed at the study of the extensive Wijk bij Duurstede excavations. Since 1992, he has been involved in physical anthropological and paleopathological research, including forensic anthropological work in the former Republic of Yugoslavia. His research interests focus on the combination of standard physical anthropological methods and bioarchaeological techniques (isotope analysis; ancient DNA; microCT) to study the demography and living conditions of ancient populations. Recent work includes the study of Late Ceramic populations in the Caribbean, human remains from a Late Neolithic burial mound, and the cemeteries of the early medieval port of trade at Dorestad.

Reniel Rodríguez Ramos is assistant professor in the Social Sciences Program at the Universidad de Puerto Rico, Recinto de Utuado, where he directs the División de Investigaciones Arqueológicas de la Montaña. He is also currently conducting postdoctoral research in the Faculty of Archaeology of Leiden University. His main research interests include the study of lithic technologies, the origins of agriculture and pottery production, and culture-contact and interaction dynamics in the Antilles and Greater Caribbean. He is the author of *Rethinking Puerto Rican Precolonial History* (University of Alabama Press) and has published articles in journals such as *Latin American Antiquity*, *Ethnohistory*, *El Caribe Arqueológico*, *Journal of Caribbean Archaeology*, and *Geoarchaeology*.

Ann H. Ross is professor of biological sciences at North Carolina State

University. She is a physical anthropologist with a subspecialty in skeletal biology and forensic anthropology. Her research focus includes developing population-specific identification standards using traditional measurement techniques and modern methods of the geometric morphometrics. She is currently involved with amassing a relational database of traditional metric and coordinate data for precontact Latin America and the Caribbean of samples dating between AD 1 and AD 1500 to examine precontact biological variation and assess patterns of migrations. She is a contributor to *Digging Deeper: Current Trends and Future Directions in Forensic Anthropology and Archaeology.*

Theodore G. Schurr has investigated the genetic prehistory of Asia and the Americas through studies of mtDNA and Y-chromosome variation in Asian, Siberian, and Native American populations for over twenty years. His laboratory is currently analyzing genetic diversity in indigenous populations of the Altai-Sayan region, archaeological populations from the Lake Baikal region, and Native American populations from Argentina. In addition, he is working with Native American populations from North America and Mexico as part of the Genographic Project. Schurr is currently an associate professor in the Department of Anthropology and a consulting curator of the American and Physical Anthropology Sections of the University of Pennsylvania Museum of Archaeology and Anthropology.

Joshua M. Torres is a Ph.D. candidate in anthropology at the University of Florida. He received his M.A. in archaeology from the University of Colorado in 2001 and has worked as an archaeologist and Geographical Information Systems (GIS) specialist in both the public and private sectors. Torres's research interests are focused on GIS applications in archaeology, social landscapes, settlement patterns, the archaeology of communities, and the development of incipient political institutions. He is currently a project archaeologist for Southeastern Archaeological Research (SEARCH), Inc.

Douglas H. Ubelaker is a senior scientist and curator of physical anthropology at the Smithsonian Institution's National Museum of Natural History in Washington, D.C. Since 1977, he has served as the primary consultant in forensic anthropology to the Federal Bureau of Investigation Laboratory in Quantico, Virginia. He has published extensively on a range of issues in forensic anthropology and human skeletal biology, including analysis of archaeologically recovered human remains from throughout the Americas.

# Index

*Tracing Childhood: Bioarchaeological Investigations of Early Lives in Antiquity*, edited by Jennifer L. Thompson, Marta P. Alfonso-Durruty, and John J. Crandall (2014)

*The Bioarchaeology of Classical Kamarina: Life and Death in Greek Sicily*, by Carrie L. Sulosky Weaver (2015)

*Victims of Ireland's Great Famine: The Bioarchaeology of Mass Burials at Kilkenny Union Workhouse*, by Jonny Geber (2016)

*Colonized Bodies, Worlds Transformed: Toward a Global Bioarchaeology of Contact and Colonialism*, edited by Melissa S. Murphy and Haagen D. Klaus (2017)

*Bones of Complexity: Bioarchaeological Case Studies of Social Organization and Skeletal Biology*, edited by Haagen D. Klaus, Amanda R. Harvey, and Mark N. Cohen (2017)

*A World View of Bioculturally Modified Teeth*, edited by Scott E. Burnett and Joel D. Irish (2017)

www.ingramcontent.com/pod-product-compliance
Lightning Source LLC
Chambersburg PA
CBHW020846270326
41928CB00006B/574